AFRICAN PHILOSOPHY, CULTURE,

AND

TRADITIONAL MEDICINE

AFRICAN PHILOSOPHY, CULTURE,

AND

TRADITIONAL MEDICINE

by

M. Akin Makinde

Ohio University Center for International Studies
Monographs in International Studies

Africa Series Number 53
Athens, Ohio 1988

© Copyright 1988 by the
Center for International Studies
Ohio University

Second printing 1989

Library of Congress Cataloging-in-Publication Data

Makinde, M. Akin
    African philosophy, culture, and traditional medicine/ by M. Akin
Makinde
    p. cm. (Monographs in international studies. Africa series ;
no. 53)
    Bibliography. p.
    Includes index.
    ISBN 0-89680-152-7
1. Philosophy, African. 2. Africa, Sub-Saharan--Civilization.
3. Medicine, Primitive--Africa, Sub-Saharan.   I. Title
II. Series.
B5375.M35   1988
199'6--dc19   88-15680
                                           CIP

ISBN 0-89680-152-7

To my father
of blessed memory

CONTENTS

ILLUSTRATIONS

1. Olugbohun, a special kind of Ase as prepared in the horn of a bull and partially wrapped with white cloth. It is one of the most powerful.

2. Ase Iwo Igala, prepared in the horn of Igala, a species of deer.

3. Assorted kinds of empty horns used for the preparation of Ase of various kinds. They range from the horns of antelope to those of buffalos.

FOREWORD

Dr. M. A. Makinde's book, portions of which were presented during his Fulbright lecture at Ohio University in 1984, is the first full-scale attempt to explicate the theoretical foundations of traditional African medicine, especially among the Yoruba of Nigeria. As such, it is a very welcome addition to the scholarly literature of contemporary African studies. Dr. Makinde's qualifications are eminently suited to the task he has set himself in this book. He grew up in the Yoruba area of Nigeria, and studied, taught, and published for many years in Philosophy of Science in North America and Awolowo University (formerly University of Ife), Nigeria. Dr. Makinde is also the son of one of Nigeria's most prominent traditional medical practitioners.

In addition to his work in Philosophy of Science, Dr. Makinde is well known in West African philosophical circles as a leader in the fight to have African traditional thought recognized as a distinct kind of philosophy. The effort to assert the independence of West African countries has taken many different forms. Besides the more obvious movements for political independence, which culminated in the 1960s, there is also the ongoing struggle for economic independence. Perhaps least recognized, however, is the effort to achieve what we may call "cultural independence," independence, that is, from unfavorable comparisons of indigenous African cultural achievements with those of the European West. So long as Western versions of art, literature, and history are considered the paradigm models, then comparable African contributions must always be judged derivative and inferior, or altogether absent. The move toward "cultural independence" is the attempt to judge African artistic and literary achievements on an independent and equal footing, judging African contributions to the arts by African standards. But this movement, begun in the 1950s, has not proceeded at the same pace in every area of cultural achievement. First, and most readily accepted, was the attempt to "liberate" African history from its colonial past. Then came the largely successful attempt to rescue African art from a subservient relation to a more "civilized" Western art and to appreciate and judge African art on its own terms and by its own standards. But similar efforts in the area of African Philosophy proved more difficult, with some African philosophers arguing that traditional African thought systems were not sufficiently like Western Philosophy to merit being properly called "philosophy" and some, led by Dr.

xi

Makinde, arguing that they were philosophical in an African sense and that that was the conception most appropriate for judging African traditional thought.

That debate has been raging for almost twenty years now. What is radically new in this book of Makinde is the attempt to carry the movement toward "cultural independence" into the area of medicine as practiced for thousands of years, indigenous forms of African medicine, supported by theoretical systems of intellectual thought very different from those of modern science which undergird modern Western medical practice. What are we to make of these traditional African medical practices and their accompanying theories? Do we judge them by the standard of Western medicine, or do we judge them in a parallel but independent fashion? From a practical point of view, are these medical practices safe and effective? Should they be used today in place of Western medicine? As a matter of fact, in Nigeria today, as in other African countries, the government supports a two-tiered approach to medical care, encouraging both traditional medicine (mainly in the rural areas) and Western medicine (mainly in modern urban hospitals).

Above all, who should make such judgments? Most of the debate on African culture over the past two hundred years has been conducted by Europeans for Europeans. What is refreshingly new in Dr. Makinde's approach is that he examines the question from an African perspective. In his defense of the effectiveness of traditional African medical practices and of the traditional African theories underlying these practices, Dr. Makinde's conclusions will seem radical to many readers outside of Africa and hard for them to swallow. But this is a typically European response which we have heard many times before and scarcely need to be reminded of again. If we are truly to begin to talk to Africans and not just talk about them, we need to hear the African side of the story from the African point of view. And this is what Dr. Makinde gives us in his book--in educated English, and couched in terms of contemporary philosophy of science, but nonetheless decidedly from an African point of view. Of course, we may disagree with him, and if we do we must engage him directly in debate, but at least we must hear him, telling his own story in his own way.

Gene Blocker

PREFACE

The present work has behind it a long history of careful thought and challenge. In February 1969 I was a member of the Residence Exchange from University College, Toronto, to the University of Chicago. I had been admitted for postgraduate study in the Department of Philosophy where I had hoped to work with Professor Vere Chappell who, disappointedly, had to leave for the University of Massachusetts in the fall of 1969. However, I had a meeting with Professor Chappell in his office. Our discussion revealed to me what could then be called the poverty of philosophy in Africa. I say poverty because the discipline of philosophy was yet to find its feet in African institutions of higher learning. Judging from economic conditions in African countries Professor Chappell thought that philosophy might be a luxury and wondered why I had chosen philosophy as an academic career. Perhaps I intended to teach the subject outside Nigeria. This was the first time the thought came to my mind that I might have been doing the wrong thing. As it happened, my inquiry showed that philosophy was not a popular course in the Nigerian universities. If mentioned at all, the subject was in connection with religious studies. The University of Ibadan, the premier academic institution in Nigeria, had no department of philosophy. The then Vice-Chancellor of the University, Professor T. A. Lambo, and his deputy, Professor Oyenuga, happened to be visiting in Canada during the 1969-70 academic year. From my discussion with them it appeared that philosophy was not yet a part of their institution's priorities.

At the University of Ife (now Obafemi Awolowo University) there was a Department of Religious Studies and Philosophy, with a great emphasis on the former. The then Vice-Chancellor of the University, the late Professor H. A. Oluwasanmi, seemed to have appreciated the importance of philosophy in the building of a nation and clearly understood the role it had played in Western civilization. He did not see it as a luxury. In fact, he saw philosophy as an important discipline which is relevant to the development of ideas, critical and theoretical thinking in all academic fields, be it in science, medicine, education, law, economics, politics, sociology, human conduct, or ways of life. If philosophy had helped in the development of Western education, politics, law, and human conduct, its study by African nations could not be seen as mere luxury: it was a necessity. Thus, when Professor Oluwasanmi visited Ottawa, Canada in the summer of

1970, one telephone call between us assured my employment as a lecturer in philosophy at Ife on the completion of my dissertation in the University of Toronto in 1974. In the 1975-76 academic year, philosophy broke its traditional tie with religious studies at the University of Ife and became an autonomous department under the headship of Professor J. O. Sodipo, now the Vice-Chancellor of Ogun State University, Ago-Iwoye, in Ogun State of Nigeria.

During my interview for appointment at Ife, the most important question that was raised was how I thought I could adapt my training in Western philosophy to African conditions, especially as my dissertation on logic and scientific method was somewhat technical. I must confess that this question has been the overriding influence on my present work. Between 1974 and 1977 I looked at philosophy purely from the Western analytic approach and almost wrote off the possibility of doing any philosophy--Oriental or African--other than that in which I had received during nine years of training, that is Western analytic philosophy. But as it dawned on me that there was more to philosophy than conceptually analytic rigor, and as I carefully examined and compared some African metaphysical philosophy with some of those with which I was already familiar in Western philosophy, I became more liberal and better prepared to probe into some philosophical problems in what was then known as African traditional thought. The best encouragement I had was through various discussions with my late father, Daniel Kayode Makinde. From him I learned quite a lot about African thought that was of great philosophical importance. The recognition of them as philosophically interesting issues was due to my own philosophical training, a situation that would have been different were I to have been trained in anthropology, sociology, or ethnography.

My interest in traditional medicine was also greatly influenced by my late father who was a well-known traditional healer in Ekiti Division of Ondo State, Nigeria. My idea to do some work on traditional medicine began when he was still alive. He was particularly worried that his knowledge of traditional medicine, like his predecessors', would be lost unless we wrote it down. I had only begun to document some of his stock of knowledge when he died suddenly on 19 February 1981. With the assistance of my mother, the close associates of my father in the same profession, and whatever documents I could lay my hands upon, I went into the field to do some research on traditional medicine to supplement my research in African philosophy. Because I am not a trained medical practitioner, I approached the subject from the cultural and philosophical points of view.

In this connection, I must express my gratitude to the University of Ife's Central Research Committee for providing me with the funds with which I carried out the first phase of my research under the title "African Philosophy, Culture and Traditional Medicine" for the year 1982-83. I also acknowledge my indebtedness to the following informants: my mother, Mrs. Dorcas Ibirinlade Makinde, Chief J. A. Lambo, National President of the Nigerian Association of Medical Herbalists, Mr. Ifayemi Akinrinde Eleburuibon of Osogbo, a well-known Ifa priest and Director of the program "Ifa Olokun A sorod'ayo" at the Nigerian Television Authority (NTA) Ibadan, Nigeria, Chief Onibedo of Imesi-Ekiti, Mr. Adesoro Alajagun of Itamerin-Osu, via Ilesa and Chief Awoyefa, the Jolofinpe of Ile-Ife. Apart from my father, my greatest indebtedness goes to his long associate in the field of traditional medicine, Chief James A. Abiola, the Eisikin of Ayegunle-Ekiti, and Professor Wande Abimbola, the internationally known scholar on Ifa and Traditional Cultures, formerly head of the Department of African Languages and Literatures, University of Ife, and now the Vice-Chancellor of the same institution. Finally, I wish to express my gratitude to my younger brother and Research Assistant Amos Morakinyo Makinde, a postgraduate student in the Department of Botany, University of Ife. A portion of my research findings, particularly in the treatment of mental illness, had already been presented in a paper, "Cultural and Philosophical Dimensions of Neuro-Medical Sciences," at The 1982 Joint Conference of the Association of Psychiatrists in Nigeria, the African Psychiatric Association and the World Federation for Mental Health held at the University of Ife, Ile-Ife, Nigeria, from 22-25 September 1982.

My Fulbright Program could be seen as a continuation of the research which started at the University of Ife in 1982 under the same title of the present work. Between September 1983 and June 1984 when I was a visiting Fulbright scholar in the Department of Philosophy at the Ohio University, Athens, Ohio in the United States, I read several papers on Western and African Philosophies both at the departmental colloquia and at international conferences. It soon occurred to me that in order to leave something of good memory to the Council of International Exchange of Scholars (CIES) which handled the Fulbright Program, and my host institution, Ohio University and particularly its Department of Philosophy where I had enjoyed one of the best facilities available for my research activities, I had to do something special. That something was my founding of an annual lecture to be known as the Fulbright Hays Lecture. The purpose of the Fulbright Hays Lecture is for the Fulbright scholar to share his research work with the university community, in a well-organized public lecture, towards the end of his Fulbright program in his host institution, whether in America, Europe, Asia, or Africa. The present work is a revised and enlarged version of the lecture I gave on 2 May 1984 to begin the series.

In the development of the First Fulbright Hays Lecture in the United States particular mention must be made of the cooperation I received from the President of Ohio University, Doctor Charles Ping, the Chairman of the Philosophy Department, Professor Warren Ruchti, the Associate Provost of International Studies, Professor Felix Gagliano, who later became the Chairman of the Ad-hoc Committee set up by the University to plan the logistics for its first Fulbright Lecture, Professors Gifford Doxsee, Director of African Studies and Vattel Rose, Director of Afro-American Studies program, both of whom were also members of the ad-hoc committee (in addition to the Chairman of the Philosophy Department and myself). I would also like to express my gratitude to Cassandra Pyle, Director of the Council for International Exchange of Scholars and Linda Rhoad, her executive associate and my program officer, and also Senator William J. Fulbright, all in Washington, D.C., for their enthusiastic responses to the University on my initiation of the Fulbright Hays Lecture. I was particularly delighted at the congratulatory messages sent to me and the Ohio University by the Director of CIES on the occasion of this Lecture. In addition, I owe the CIES an immense gratitude for extending my Fulbright Award from May to August 1984, on the recommendations of the Chairman, Department of Philosophy, Directors of African Studies and Afro-American studies programs, Ohio University, and Assefa Mehretu, Assistant Director, African studies Center, Michigan State University, East Lansing, Michigan. I must also express my gratitude to Chief Obafemi Awolowo of Nigeria for sending me several copies of his books on social and political philosophy during my preparation for the Fulbright Hays Lecture.

Next I think I should show my gratitude to the City of Athens for its very quiet and distinctly academic environments, free at least from the hurly burly of the cities of New York, Los Angeles, Chicago, and Philadelphia. I have profited a great deal from my discussions with academic colleagues at Ohio University, and in this regard special mention must be made of Nader Chokr, Professors Gene Blocker, Stanley Grean, and Algis Mickunas of the Philosophy Department. My thanks also go to Professor Richard H. Popkin, of the Department of Philosophy, Washington University, St. Louis, Missouri, for his encouragement and advice on a work of this nature. It will be difficult for me to express enough gratitude to Alice Donohoe, the secretary of the Philosophy Department, Ohio University, and her assistant, Harriett Lang for their superb secretarial assistance during the period of my stay from 14 September 1984 to 30 June 1984. I owed my sanity and peace of mind during this period to my loving wife, Taiwo, and my three little sons, Akinola, Olumide and Kayode. Their presence removed all fears and anxieties that I would have had about their conditions had they been out of easy reach. My thanks also go to my wife for typing the final draft of the revised and enlarged manuscript of this work. Finally, I thank the Department of Philosophy, Michigan State University for providing me with office space and secretarial assistance during my short stay at that University.

The present work, being an outcome of a lecture, the very first of its kind within the Fulbright programs, is meant to serve two distinct purposes. First, as a gratitude for all the things said above, and as an academic output deriving mainly from the topic of my research. But it is within the framework of the latter that I accordingly designate this work as "African Philosophy, Culture and Traditional Medicine." The present work, an important portion of a larger enterprise, is meant to provoke as well as stimulate further thought about the need to develop as well as integrate some aspects of African traditional culture, thought, and medicine with those of Western culture, thought and medicine, especially in this age when technology is moving the world into a convergency of cultures. The world would not be a worse place to live in if through proper integration of philosophies, cultures, and medical knowledge and practices we come to know each other's way of life and thinking. Therefore, this book is written with the modest hope that it would at least enhance further discussions on a very important issue that seems perfectly consistent with the very objective of the Fulbright Hays programs.

Athens, Ohio, 1984

Chapter 1

INTRODUCTION

Modern philosophy and medicine have been dominated by empiricism. European culture in the Middle Ages was dominated by religion, belief in the supernatural, and the Divine Right of Kings. Nearly all the philosophers of that age were associated with scholastic metaphysics and dealt mainly with scholastic philosophy and scholastic theology as inspired by their relations with Christian dogmas. Scholasticism, therefore, was simply philosophy placed under the power of guidance of Christian Catholic theology.[1] It championed the so-called traditional proof for the existence of God as the first cause or principle of existence. In the Western philosophical tradition, the belief in God and immortality could be traced to Pythagoras and Plato both of whom lived before Christ. But Pythagoras, from whom Plato borrowed the idea of the immortality of the soul, was known to have travelled to North Africa, notably Egypt, from whence he learned of the idea. Later, an African philosopher, St. Augustine, speculated about the soul which is also central to all religious belief.[2] Philosophical speculation about the existence of God and the human soul had existed both in Africa and in the Western world, and speculation about it is as old as philosophy itself. It was one of the earliest curiosities of man in his attempt to understand the universe and the human person. The idea of immortality of the soul led to a distinction between mind and body, and it is from this dualism that much of modern philosophical thinking arose.[3] Even the general speculation about God and the origin of the universe remains a strong issue in modern philosophical debates.[4] While some of these issues probably originated from Africa and got their way to the Western world through the ancient Greek philosophers the absence of written materials made it difficult to document African contributions in philosophical speculations.[5]

The transition from philosophical speculation to empiricism was only notable in Europe, particularly in Great Britain. In the early modern period a more critical attitude invaded Great Britain and consequently people's beliefs in God and the Divine Right of Kings. Non-observable entities were subjected to questioning. The sacrosanct nature of monarchy was eroded first by the empirical philosophy of Francis Bacon (1561-1626) and later by those of Thomas Hobbes (1588-1679) and John Locke

1

(1632-1704).[6] There was a radical change in culture, especially in British societies. Scholastic metaphysics which drew its inspiration from the Aristotelian philosophy was replaced by empiricism. The center of attraction became nature itself. This was reflected in Wordsworth's phrase "let nature be thy teacher." It was a spirit typical of Bacon's philosophical writings. This cultural revolution continued and came to its height during the Victorian era in nineteenth century England. In fact, most Englishmen agreed to the characterization of the Victorian period as an age of transition in British philosophy and culture.

> The legacy of Francis Bacon (says George Basalla, etc.,) haunted the British reflections upon scientific method. "Man", wrote Bacon, "being the servant and interpreter of nature, can do and understand so much only as he observed in fact or in thought of the course of nature." Bacon's path led to power over nature, power to be gained by observing nature and thus learning her ways. Generations had sought truth by fleeing from the senses and taking refuge only by pure reason. Bacon's method inspired, however, to derive these axioms "from the senses and particulars, rising by a gradual and unbroken ascent" and gaining the "most general axioms last of all."[7]

The appeal to sensory experience and empirical method advocated by Bacon thus effected a remarkable change in people's beliefs and perceptions about reality, especially about scientific knowledge. So great was the Baconian spirit of philosophy and its effect on cultural, social, and scientific change that in the nineteenth century there arose a great awareness about the industrial implications of science as scientific applications began to contribute to the wealth of Great Britain. Francis Bacon whom Karl Marx rightly described as "the real founder of English Materialism and all modern experimental science," and for whom "natural science was true and physics based on perception was the most excellent part of natural science," has thus opened the way to the understanding of nature, to the extent to which science seemed to have promised man more powerful control over nature.[8] Science thus became the arbiter of knowledge. The spirit was Baconian Philosophy and the method the Newtonian experimentalism as construed in the Baconian tradition of inductive science.

In recalling her Victorian childhood education, Beatrice Webb had much to say about the general assumption of that era of scientific advancement as affected by change in belief, culture, and philosophy. According to Webb, the general belief in that period was that physical science would solve all problems, be they social, political, medical, or ethical.[9] It was the same spirit that led John Stuart Mill, the nineteenth century British philosopher, to advocate the application of the method which had

2

been so successful in the physical sciences to the moral, social, political, and economic problems of his time.[10] The practical values of science were stressed vigorously and proclaimed so widely that Victorian technological society looked upon the utility of science as the nation's scientific maturity. As science continued to advance, it conquered diverse phenomena. Under these circumstances "it was not difficult for the confident Victorian scientist to extrapolate to that time in the future when the methods of science would be applicable to every aspect of human endevour."[11] The twentieth century Western technological culture, with all its good and evil effects on man and society, seems to have made good the "confident Victorian" scientist's projection. Western technology has indeed led to the improvement of the lot of human beings, but it has done this at the expense of worsening the general human condition. It thus appears that philosophical or scientific materialism which has led to scientific progress has also led to a paradox of technological progress. In recent years, the evil application of the monster called technology has created all sorts of problems for human societies including the effects of the computer revolution on the human labor force and self-esteem, the diabolical use of electronic surveillance for the invasion of privacy, the exploitation and dehumanization of human beings as soulless automata or material objects without souls or minds, complete alteration of the human mind in the form of mind-control and, the most dreadful of all, the daily fears and anxiety about nuclear war and the destruction of the human race and civilization.[12] These are the sad effects of technological progress whose foundation was laid by the empiricism and materialism of Francis Bacon's philosophy.

In contrast to the above, there has been little or no corresponding change in African culture and philosophy. There is no evidence to suggest that any serious change had occurred in African metaphysical beliefs about God to the extent in which empiricism and scientific method might have been advocated by African thinkers. In this connection, one may be tempted to agree with Reverend John Mbiti who, in the twentieth century, argues that Africans live in a religious universe, with God as the ultimate explanation of all things and events, and his other thesis that, as far as time is concerned, Africans have no concept of the future.[13] As this latter position has been refuted both from the points of view of the use of language and anticipation of future events among Africans, it is needless to discuss the issue any further.[14] But the other opinion of Mbiti appears disturbing as arguments could be found in its support. Although Mbiti's position suffers from overt generalization, the bulk of African philosophy seems to hover around metaphysical thoughts, most of which derive from belief in God and his heavenly or earthly agents. This belief has militated against a scientific notion of cause, especially as many Africans always attribute causes of events to God or some spiritual forces.[15]

Surely such metaphysical beliefs stand in the way of empirical science. Be this as it may, it is not true to say that all Africans live in a religious or spiritual universe, as though Africans uncritically hold one philosophical belief. Although there is no record of an African Bacon, it was perfectly conceivable that many African thinkers were empiricists who saw a need for learning from nature. The problem is that there were no identifiable figures in African philosophy until quite recently. The whole issue, therefore, seems to be clouded by a lack of written philosophical treatises or speculations by African thinkers. This was due more to the absence of writing than to the poverty of philosophical ideas.

Apart from their belief in God and other spiritual entities, some aspects of African traditional thought are quite comparable to Western philosophical thought.[16] The difference between African belief in God and Western belief does not rest on evidence adduced for such a belief but on abstract argumentation or proof for His existence. Thus while some Africans believe in God and might have had the idea of ontological, cosmological, and teleological arguments, they do not, philosophically, attempt reasoned arguments to prove their belief. But some, like the Nigerian Tai Solarin, is well-known for his skeptical arguments concerning the existence of God, particularly from an empirical point of view, that is, the argument from Evil.[17] The Yoruba theory of immortality of the soul, for instance, is similar to Plato's.[18]

But perhaps the most important aspect of African philosophical thought is the concept of a person. Some work has been done in this area by African philosophers, including the present writer.[19] The dualism of mind and body which remains a perennial issue in Western philosophy could also be seen as an important issue in African philosophy. The essential difference, of course, is the African rejection of the Western reduction of mind to matter, thus rejecting philosophical or scientific materialism on which Western science and technology is based. Although such a rejection could be seen as antiscience, and probably responsible for the poverty of scientific ideas and achievements in Africa, it certainly could be seen also as a possible antidote against dehumanization or treatment of human beings as soulless automata, capable of being scientifically manipulated like atomic particles. The African belief that human beings have souls has for several centuries created the feeling of respect for personhood. Respect for a person continues even after his death, since it is generally believed by many Africans that, after his death, a person's soul lives on in the form of a spiritual existence. Thus people say nice things about, and may even worship, the spirits of their ancestors.[20] Although such beliefs may be scientifically unsound, they certainly are philosophically and morally relevant to the development of sane societies. They have the merit of avoiding the problems created by science and technology in modern societies where the concept of a person and

4

μ p. 51 for more on Ifa + Ori

respect for personhood and human values have lost much of their meaning.

This seems to show, therefore, that a philosophy based solely on empiricism or materialism is inadequate for the understanding of the meaning and purpose of life, and the improvement of the general condition of people as opposed to their material lot on earth. It also shows that science by itself is not the measure of all things, for a strictly scientific view of life puts human beings under the control of science and technology and this in turn reduces them to inhuman and inhumane machines. Therefore, in spite of its lack of scientific value, African philosophy has something to contribute to decoding the excesses and abuses of science, at least when the general condition of human beings and their relations to their fellows are concerned. It is from this point of view that I see African philosophy as worthy of study, in spite of its immaterialism and lack of conceptual rigor. It can at least provide suitable grounds for a general review of what, properly speaking, should be the goal of philosophy or even of science.

Ifa, Knowledge, and Probability

A great deal of African philosophy has its roots in African cultural beliefs, some of which are not worth courting. Some of these beliefs may be regarded as outmoded in the twentieth century world and so ought to be forgotten. Others may be seen as so fundamental to African heritage and values that they need to be preserved, revised, or improved. There may be yet others so controversial as to demand a more critical examination, analysis, or refinement here and there in order to make them consistent with our modern beliefs. In these cases it is either that the particular cultural beliefs at stake contradict our present beliefs, so that a rejection of earlier beliefs becomes inevitable in the light of our modern experience, or they are not inconsistent with our present beliefs, but nevertheless require some further analyses by which some possible refinements or conceptual modifications might be achieved.

In this connection we shall single out the role of Ifa in Yoruba thought. As I see it, the most prominent and certainly the most dominant aspect of Yoruba culture is associated with the Ifa literary corpus. Although Ifa is not a philosophy, it has in it a great stock of ideas that generate various philosophical issues, including metaphysics, ethics, epistemology, and science, of which the most developed is traditional medical science. Several works on Ifa by Professor Wande Abimbola have paved the way for a better understanding of its importance and relevance to Yoruba philosophical thought.[21] It is from the learning of Ifa that we come to understand the Yoruba concept of a person and of immortality. Of particular interest is the Yoruba concept of ori and its relation to human destiny. In modern times the present writer has given a philosophical analysis of this concept. Such

5

analysis reveals the necessity for its conceptual modification in order to disclose its relation to, and consistency with, other concepts such as human destiny, human personality, freedom, and responsibility.[22]

Ifa, which is known as a repository of knowledge or infinite source of knowledge (imo aimo tan), is in possession of knowledge consisting of several branches: science of nature (physics), animals (biology), plants (botany), oral incantations (ofo), divination (prediction), medicinal plants (herbalism), and all the sciences associated with healing diseases (medicine).[23] It is generally known that the issue of Ifa divination presents a prima facie difficulty for empirical scientists and scientifically-minded philosophers of the Western world, especially as Ifa divination is used for prediction as well as diagnosis in African traditional healing. While an African may find the Ifa divination a workable system, a person from a Western culture may dismiss it as mere superstition insofar as he has no antecedent experience by which his belief in Ifa divination could be sustained. This shows how experience can shape a people's belief in a particular culture.

Surely, in every culture, there exists a strong desire to know about the future, that is to be able to make reliable predictions. In Western cultures, scientific development has increased hopes for reliable predictions. But Western science has a method, and scientific theories can be read in textbooks. If in African cultures there exists a method or means by which reliable predictions can be made, we shall call it scientific. But predictions in this sense shall not be restricted to those governed solely by Western scientific theories. There are other cultures which have tried, and quite successfully too, to determine future events by some other means than those open to empirical investigations. This, of course, may lead one to suggest that while the former could be regarded as "open" science, the latter could be regarded as "occult" or "restricted" science. In making this distinction we may be suggesting that science is not, without remainder, empirical, just as it is often argued that logic is not all purely formal.[24] When a particular view of science is not open to empirical investigations, some empirical scientists call it all sorts of names: metaphysical nonsense, mysticism, occultism, fetishism, spiritualism, magic (black or white). But if our contention that not everything under the sun that is called science is empirical without remainder is tenable (this contention is highly controversial), then the calling of non-empirical science with all sorts of pejorative names as indicated above is without any conclusive justification.

In Africa, and particularly among the Yoruba, the most potent source of having insight into the future is the Ifa oracle. But Western civilization has led to neglect of Ifa, thus creating an intellectual stupor against its conscious development. In Yoruba society Ifa is commonly used by experts to

6

foretell what is likely to happen, given that certain conditions remain as they are. This means that while Ifa can predict what is likely to happen if certain conditions remain unchanged, it also can prevent an event from occurring, granted that certain other conditions are either prevented or eliminated in the manner suggested by the Ifa oracle. In this case Ifa serves a dual purpose. It foretells by warning, and provides solutions to anticipated events or problems. In many cases, Ifa appeared to have worked. That may serve as its own justification.

It is very important to point out here that an Ifa priest needs a long period of training in order to acquire his knowledge. Hence, just as not anybody can be a theoretical physicist or a surgeon, not just anybody can be an Ifa priest. An Ifa priest may require an even longer period of training to go through the manipulation of all the 256 steps known as "Odu Ifa." There are two categories of the 256 odus, the major 16 and the minor 240. In order to become an Ifa priest one has to study properly the literature of each odu, one by one, with a minimum of 16 versus or stories in each odu. Altogether, one is required to master 4,096 (16 x 256) odu before one graduates as an Ifa priest. In the Yoruba culture the period of training for the Ifa priesthood may be from the time one is a child to the time one becomes an adult. This is very common in cases where the priesthood is to be inherited from one's parents. In other cases where it is not inherited from one's parents, young men and women serve long periods of training and apprenticeship under reputable Ifa priests. We rarely find Ifa priests who are less than thirty years old. Their average age is between forty and fifty. It appears, then, that Ifa oracular power is not easy to acquire. This situation seems to me comparable to Plato's recommendations concerning the period between the study of philosophy or dialectics and its application by a philosopher king. Apart from the preliminary studies in earlier years, Plato gave between the age of thirty and fifty years for the study of philosophy and the acquisition of the necessary experience and practice to become a philosopher king.[25] This possible comparison between the Yoruba Ifa priesthood and the period of training it requires, and Plato's views about philosophy or dialectics and philosopher kings, is likely to suggest to us the important role of Ifa in Yoruba epistemology. I say this because of the role Plato's notions of dialectics and the philosopher king play in the ladder of his theory of knowledge.

At present, the concept of Ifa is receiving some critical analysis, and being given philosophical and probabilistic treatment, particularly at the University of Ife. From the point of view of epistemology it may very well supply us with some ways of knowing, whether or not its general method is open to empirical investigation. After all, there are both empirical and non-empirical (rationalist) ways of knowing. The kind of epistemology we admit depends on our philosophical temperament. But with respect to science (not the so-called exact sciences like logic and mathematics), we can at least suggest that the kind of knowledge philosophers usually talk about reflect probability rather than certainty. Therefore, we shall confine ourselves to the fact that the whole of the Ifa literary corpus can be given a probalistic interpretation. What is more, it can be developed into science. I believe that philosophers can provide the theoretical background on which a probalistic interpretation of Ifa oracles can be worked out. We shall call this an African science of the future.

For the above reasons we may assume that the Ifa oracle bears a resemblance to Laplace's omniscient intelligence which would be in possession of causal laws that would allow for the prediction of the whole future of the world from the knowledge of the present state.[26] For the time being, the question as to whether or not the existence of such an omniscient intelligence is possible is not important. The only way in which anything can be proved to be non-existent is to prove that its existence is impossible. A centaur, for example, is non-existent on this earth because what we call a centaur is impossible under conditions of terrestrial physiology. And there are certain gods, such as the gods of Homer's Olympus, whose existence may be shown to be impossible under natural laws. But unless we can assume that these natural laws not only obtain in, but, in fact, rule the entire universe, we cannot prove that even the gods of Homer or the Yoruba gods of iron (Ogun) and thunder (Sango) are impossible everywhere. In the same way the non-existence of an omniscient intelligence cannot be proved. It is sufficient to allow that every culture has either intuitive or pragmatic reasons for believing in these things. With this little digression we shall count ourselves as being in a safe position to compare Ifa with Laplace's hypothetical omniscient intelligence.

Ifa, like omniscient intelligence, would be in possession of causal laws that would enable those who thoroughly understand its operations to make predictions about the world, given the background of its present state. Just as trained scientists would only have to consult with this intelligence for their predictions, so also would the trained Ifa priests have to consult with the same intelligence, known as Ifa, for their own predictions. The only difference between the scientist and the Ifa priest would either be a question of terminology or the difference in their methods of consultation and investigation. But above all, the essential difference would be seen only in the openness of the one and the secret nature of the other. To argue that the methods derived from the latter are not displayed openly in scientific textbooks does not mean that they cannot be so displayed in the future. In the very near future, the methodology of the Ifa oracular system may turn out to be an important discovery. The only thing that needs to be said is that if the hypothesis of an omniscient intelligence were true, then such knowledge that could be in the possession of scientists and Ifa priests may seem wonderful to have, and may very well be the kind of knowledge the magician Paracelsus has in mind in his <u>Sagacious Philosophy</u> (Philosophia sagaz) where he contends that "magus" is in command of the forces of nature. The "magus," according to Bombastus Paracelsus, is the wise man to whom Nature has taught her secrets. He knows the signs which reveal her powers.[27] In fact, the possession of such knowledge, which may be the real thing behind the theories of scientists as well as the manipulations of Ifa priests, may also make it rather difficult for us to know the real difference between scientific predictions and the predictions of the Ifa priests on the one hand, and either science or magic on the other.

However, because of the critical nature of Laplace's society and culture, his speculation about an ideal of perfect knowledge that could be gained through consultations with an omniscient intelligence was reduced, at its best, to mere probability. This is to say that in the absence of a postulated ideal, such as an omniscient intelligence or perfect knowledge, as finite beings we must content ourselves with "partial knowledge, mingled with doubts, and producing ignorance."[28] The measure of such knowledge, according to William Stanley Jevons, is the Laplacian Theory of Probability. Since the idea of perfect knowledge as attributed to an omniscient intelligence out of critical consideration may not be available to both scientists and Ifa priests (because it is only an ideal to be approximated), both the scientists and Ifa priests would have to content themselves with probability. In this connection, a probalistic interpretation of Ifa is possible. When this is done, the criteria for the acceptance of the predictions of both Ifa and science would depend on their reliability and the extent to which each, as a guide to life, approximates to the ideal. This is to say that, if properly analyzed, the difference between the scientist's predictions and those of the Ifa priest or Babalawo may not be one in kind, but in the methodology and efficacy of each. And to the extent to which our Babalawos never claim that their predictions about events are known with certainty, so Ifa oracles claim no more than science claims about our knowledge of the world, that is, probability.

In support of our claim that the difference between the scientists' predictions and those of the Babalawos may not be in kind but in methodology, we shall examine an interesting philosophical point about predictions in general. Let us imagine that a scientist in the Western culture makes 100 predictions about some state of affairs over the years, and ninety of them come out true. Now, let us imagine that a Babalawo (or even a soothsayer), after consulting the Ifa Oracle, makes 100 predictions about the same state of affairs over the same period, and ninety of these predictions also come out true. Irrespective of which of these ninety predictions come out true from either side, and also irrespective of methodology, there arises an interesting philosophical question. Which of these two do we accept as the better predictions? Do we accept the scientist's or the

9

Babalawo's as the better predictions or do we simply accept one set of predictions and ignore or reject the other when both sets of predictions are ninety percent successful? Do we dismiss one as mysterious or magical and so not worthy of serious attention? Any attempt to do this would be philosophically untenable, and could be dismissed as arising from mere prejudice, cultural divergencies, different philosophical temperaments, and differences in the way individual cultures look at the world and solve their own problems. Perhaps the predictive power of Ifa divination might be acceptable to the Western world of empirical science if its methodological principles were clearly defined and theories on which its predictions are based are well explained in textbooks. In the absence of this, Ifa divination may appear to be a superstition, but a superstition which, nevertheless, often turns out to work quite effectively like a non-superstitious science. After all, there are varieties of reasons for accepting or rejecting theories--even scientific theories.

It is a happy thing to note that the World Health Organization has come to recognize the value of traditional medicine and practice in Africa. Given the escalating cost of medical care in the developed countries of the world and its infiltration to Africa and other less developed countries such as India and China, there now exists an awareness of the urgent need to develop systems of traditional medicine and their integration with Western, orthodox medicine and medical practice as a way of bringing medical services nearer to the majority of people in the rural areas where most people rely on traditional healing. But as we shall see later in our discussion on traditional medicine, some Western-trained doctors appear to have stood against the encouragement and development of traditional medicine for reasons best known to them. Neither do the African governments show any seriousness of purpose in their often proclaimed desire to develop indigenous systems of institutions as done in India and China.

Perhaps the greatest problem of Africa as seen in the general attitude of her philosophers and medical doctors, as well as those in other intellectual disciplines, is that of language. To this important issue I shall address myself later on, particularly to the relation between language and culture and the influence of both on philosophy and modern science. The purpose of such discussion will be to show how the adoption of foreign languages has affected African cultural systems. Such adoption has also led to a general feeling of inferiority in many intellectual endeavors since philosophers, medical doctors, scientists, and intellectuals from other disciplines are forced by language to depict their conceptions of reality in the language of foreign cultures as opposed to the language of their own cultures. This I take to be one of the most important causes of the skepticism against African philosophy and traditional medicine by some contemporary African philosophers and medical doctors trained essentially in the Western philosophical and medical traditions.

One innovation in the present work is the introduction of some historical phases in the study of African philosophy. Although it has been argued by some scholars that Greek philosophy was influenced by African thought, the only great figure known in African philosophy was St. Augustine (354-430 A.D.) whose philosophical writings are well known.[29] A history of African philosophy could then be traced to Pythagoras's contact with Egypt around 570 B.C., although nothing was written by Africans from that time until the time of St. Augustine. From all indications, nothing much was recorded on African philosophy before and after Christ. The situation was no better during the Middle Ages. Not even was much known in writing about African philosophy far beyond the Middle Ages. The first phase, therefore, was very uneventful, and the absence of great figures to be identified with their philosophical works, as was the case in the history of Western philosophy, actually led to some of the problems concerning African philosophy today.

It was not until the first half of the twentieth century that some colonial scholars, mainly Christian missionaries, anthropologists, and ethnographers came on the scene. Apart from their claim that African philosophy represents the group or collective thought of a people, having taken the Western standard to judge the African system of thought, most of them came out with the conclusion that African cultures and traditional thoughts were prescientific and prelogical, using the term "prelogical" to describe a kind of thought that is not free from inner contradiction. Others think that Africans live in a religious universe, so that all the activities and thoughts of the Africans can be expressed and understood from the point of view of religion. These, of course, include philosophical and scientific activities. With the emergence of the Western-trained philosophers in Africa, the views of the colonial and post-colonial scholars were subjected to criticism, especially their conceptions of African philosophy as group philosophy. This conception, which was rejected, has been described by African contemporary philosophers as ethnophilosophy.[30] Belonging to this group of colonial scholars were Lucian Levy-Bruhl, Reverend Father Placide Tempels, Reverend Father Alexis Kagame, Janheinz Jahn, Robin Horton, and Reverend John Mbiti.

Since the late 1960s, Western-trained philosophers have begun to emerge among African scholars. There are now quite a large number of them, belonging to two different linguistic groups: Anglophone and Francophone. They felt dissatisfied with a conception of philosophy as ethnophilosophy, especially when dealing with African thought. While they emerged as critics of colonial ethnophilosophers, they also began a controversy as to whether or not there is African philosophy. This debate, at present, is almost entirely negative, and there is no general agreement over the question of African philosophy. In spite of

11

this controversy, the teaching of African philosophy is gaining ground in Africa, and even in the United States, owing to the amount of literature now available. To the contemporary Anglophone philosophers belong the following: J. O. Sodipo, formerly Head of Department of Philosophy, University of Ife, Nigeria, now Vice-Chancellor (President), Ogun State University, Ago-Iwoye; M. Akin Makinde and an American, Barry Hallen (both at the University of Ife, Nigeria); P. O. Bodunrin (University of Ibadan, Nigeria); Kwasi Wiredu (University of Ghana, Accra); William E. Abraham (Ghana); and H. Odera Oruka (University of Nairobi, Kenya). To the contemporary Francophone philosophers belong Paulin Hountondji (University of Benin at Cotonou); Niamkey Koffi and E. Boulaga (University of Ivory Coast at Abidjan); M. Towa (University of Yaounde, Cameroon); Alassane N'Diaye; and Cheikh Auta Diop (University Dakar, Senegal). This list is by no means exhaustive of the contemporary African philosophers in Anglophone and Francophone Africa. The most notable of the contemporary African social and political philosophers are Chief Obafemi Awolowo (Nigeria), Leopold Senghor (Senegal), Julius Nyerere (Tanzania), and the late Kwame Nkruma (Ghana).

Throughout our discussion in this book particular references to African philosophy and traditional medicine will be made to some specific linguistic communities. In order to avoid overt generalization I will illustrate most of my points with Yoruba philosophical thought and system of traditional medicine. It will be found, for instance, that some of the things that obtain among the Yoruba, Igbo, or Hausa of Nigeria also obtain among the Akan or Ashanti of Ghana, but this does not mean that there are no different philosophical beliefs among the African people. My discussion on African philosophy and traditional medicine must not be seen as that of group thought or of a collective mind of the whole of Africa, as though all Africans have the same perception of reality. Three well-known British philosophers-- John Locke, Bishop George Berkeley, and David Hume--came from different geographical areas of Great Britain. Locke was an Englishman, Berkeley an Irishman, and Hume a Scot. Thus, philosophical ideas that were propagated by Locke, Berkeley, and Hume were known, not by the name of English, Irish, or Scottish philosophy, but by the name of British philosophy. And, in a more general sense, Descartes' philosophy (French) and Kant's philosophy (German) are both referred to as Continental philosophy. In its most general sense what we know as Western philosophy consists of the British, European Continental, American, and Canadian philosophies among others.

Therefore, in order to avoid an academic double standard, we should hold that the phrases "British philosophy," "German philosophy," and "American philosophy" qualify as ethnophilosophy unless we remove the words "British," "German," and "American" and substitute "Locke's," "Kant's," and "Dewey's." Since Locke's philosophy, Kant's philosophy and Dewey's philosophy do not represent the totality of the thought of the people of Great

Britain, Germany, and America respectively, the very idea of British or German or American philosophy must be dropped. If African philosophy is seen as ethnophilosophy, so must British, German, or European philosophy. As a corollary, the word "tribe," an English word of Latin origin, is applicable to any cohesive group in human societies. Thus, if we can refer to the Yoruba of Nigeria and the Akan of Ghana as tribal groups, we can say the same thing of the Irish tribe in Great Britain, the French Canadian tribe in Canada, the Italian, Hispanic, and Jewish tribes in America. Either the word "tribe" is applicable to groups of human beings with the same objectives and aspirations as in the above or it is dropped entirely from the English vocabulary. By the same token either the idea of ethnophilosophy used to describe African philosophy is dropped or it is seen as applicable to all philosophy designated by geographical names. Perhaps philosophy shall no longer be called British, German, American, Indian, Chinese, or African but qualified by names of individual philosophers, irrespective of geographical location, e.g., Bertrand Russell's philosophy, Dewey's philosophy, Kant's philosophy, Confucius's philosophy Buddha's philosophy, Awolowo's philosophy, Wiredu's or Hountondji's philosophy. We can then adopt what I shall call "ecumenical philosophy" as covering description of the philosophies of X, Y, and Z (philosophers) and of A, B, and C (nationalities).

The purpose of the above argument is to forestall any description of African philosophy and traditional medicine in pejorative terms. As Hountondji has rightly observed, when applied to African thought, the word "philosophy" is not used in its strict sense of an academic discipline. Rather,

It is used in the broadest sense of a collective and implicit world view. However, if you speak of French, British or American philosophy, you never mean "the collective world view of the French, British, German, or Americans." Now, when you get to Africa's case, there is a secret change of semantic register, a surreptitious change of meaning to the word "philosophy" so that "African philosophy" does not mean the total explicit philosophical discourse elaborated by African thinkers, but an implicit and collective world view of African peoples, as if the meaning of "philosophy" has to change with the change of its geographical application.[31]

The present writer is in perfect agreement with Hountondji's challenge of this semantic manipulation of the word "philosophy," and indeed the word "tribe," when they are applicable to African thought and people. It appears as if the study of anthropology and enthnography, which specialize in semantic manipulation of words to describe certain "primitive" societies, would never have existed had there been no Africa. But in reality, there are primitive Europeans as there are primitive Africans.[32]

What the present writer is asking for is that if philosophy described by nationalities or certain parts of them (for example, Berkeley's [Irish] philosophy which is also known as British philosophy) are not called by the name of "ethnophilosophy," that is a "collective thought" or "group mind" of a people, there is no justification for thinking that, in the case of Africa or part of it, some special terminology must be coined to describe African philosophy. From this point of view we hold that what is good for the goose is also good for the gander. Accordingly, a philosophical view from any part of Africa must be seen as an African philosophy whether or not it is different from or similar to that of other geographical areas of the continent. My discussion on some aspects of Yoruba philosophical thought as well as traditional medicine, or indeed, of the Igbos of Nigeria or Akans of Ghana must be seen, therefore, as a discussion on African philosophy and traditional medicine, and not on "ethno-philosophy" and "ethno-medicine."[33] In the final analysis, these are pejorative terms.

Chapter 2

ON PHILOSOPHY AND CULTURE

The word "culture" has a long history of definitions and interpretations. During the Victorian era culture was proclaimed to be a curtain which divided classes, religions, political parties, and even university faculties. Matthew Arnold, a great protagonist of culture, enlarged this definition of culture to include moral values. Culture was defined by Arnold as the "study of perfection" moved "by the social passion for doing good," while he saw the great aim of culture as that "of setting ourselves to ascertain what perfection is and to make it prevail." Some spoke eloquently for science in their own assumptions about culture. The more cynical minds like Richard Bright saw culture as "a smattering of the two dead languages of Greek and Latin." It has been noted that the above set of Victorian meanings of the word "culture" has persisted to this day, comprising, as it does, the practice of knowledge of the arts. Its attitudes are seen to be those of the "intellectual" or the "highbrow."[1]

In the twentieth century an anthropological dimension was added to the Victorian definitions of culture. In its widest sense culture now stands for a people's traditions, manners, customs, religious beliefs, values, and social, political, or economic organization. Culture in this sense does not refer to an individual but to a people as the individual writ large. It is not static but evolutionary.[2] Judging, therefore, from the fact that the word culture could be given different meanings and interpretations, I shall discuss my own views about the relevance of culture to a people's system of belief and, ultimately, their philosophy.

Insofar as everybody belongs to an age or culture, then to whatever school he or she may belong a philosopher is first and foremost a person of culture, a product of the education and belief of his society. If a philosopher in one culture sets a higher standard of philosophizing than some others in other cultures, it is because one culture sets a higher standard of education, belief, knowledge, moral, and social values than some other cultures, the practical end of which would be the training of people to be good members of the society. This may include the propensity for critical examination of such things as accepted beliefs, particular ways of looking at the world, acquisition and utilization of knowledge for the improvement of the material well-being as well as the general conditions of human beings as thinkers, doers, and undisputed instruments of social, political, and scientific changes.

Culture, some believe, bears some imprint of the creative power of people.[3] In this case, the imprint is that of an individual thinker in a particular culture. Since not everybody in a culture possesses a creative power, not everybody in a society is a philosopher or a scientist. While both the philosopher and the layman share a similar culture because both are products of the same culture, the philosopher is not wholly bound by his lay culture and thus has an autonomous standing for criticism. To those people who accept beliefs and norms without further criticism culture is static, not evolutionary.[4] They live in what Popper calls World Two, the world of pure subjective beliefs.[5] They would be like the amoeba which, because they lack the critical attitude of an Einstein, live and die with their beliefs.[6] This is like living in a world of unchanging beliefs. Although the philosopher is a product of his culture, his creative and argumentative power does make dramatic changes in cultural beliefs and provides innovations such as we see in great revolutionary ideas that have led to the revision or total abandonment of old ideas, culminating, so to say, in one of the most pervasive, useful but yet dehumanizing ideas of the twentieth century technological culture.[7]

Philosophy is related to culture in the sense in which a philosopher looks at the world from the point of view of the beliefs and circumstances of his life as well as those of his people and culture. Bertrand Russell said as follows:

> To understand an age or a nation we must understand its philosophy, and to understand its philosophy we must ourselves be in some degree philosophers. There is here a reciprocal causation: the circumstances of men's lives do much to determine their philosophy, but conversely, their philosophy does much to determine their circumstances.[8]

This interaction between a people's circumstances and philosophy has existed in the form of different philosophical systems throughout the history of philosophy, from the ancient through the medieval to the present world of materialism. Dominated by technological cultures, the present world is torn between two political and philosophical world views, each of which is essentially monopolistic in its conception of a better world: capitalism and communism. These are two cultures which often reflect the thought system of their people. Thus Russell said the following:

> Philosophers are both effects and causes: effects of their social circumstances and of politics and

institutions of their time; causes (if they are fortunate) of beliefs which mould the politics and institutions of later age.[9]

Philosophers, therefore, are the effects of the cultures, beliefs, and circumstances of their time; but as philosophers they could also be causes of changes in beliefs and circumstances in a culture. This has been the case throughout the history of philosophy.

No discussion on the relevance of culture to philosophy is adequate without examining its relation to language. As different cultures have different languages, so we may expect different philosophical temperaments. We do assume that two cultures speaking different languages need not have the same philosophy. Thus we have British empiricism and Continental rationalism including German idealism; the tough-minded empirical frame of mind and the tender-minded rationalist frame of mind, as William James would say.[10] Therefore, it is probable that opposing philosophical temperaments resulting from different cultural backgrounds would influence people's ideas about economic, social, political, scientific, religious, ethical, and aesthetic beliefs in accordance with the differences in cultural beliefs and philosophical temperaments. The philosophical frame of mind that is distinctly African in any important sense is yet to be established. One of our difficulties is in respect of language. And since language and cultures are closely related, beliefs and ideas in a particular culture must be reflected by its own language system.

One may suppose, like Wittgenstein in his Tractatus, that the limit of our language is the limit of our world. Since every culture has its own belief and perception of the world, it could be said that the limits of language in a culture are also the limits of the culture's perception of the world of reality. It has been suggested, for instance, that in the langauge in which human beings are familiar, not all terms have perceptual meanings.[11] Although in the Western languages perceptual references are almost always about objects, properties, or relations, it has been shown that perceptual references may also be about processes. Evidence has been given that this is probably the dominant kind of perceptual reference in some American Indian languages.[12] Since percepts do really belong to experience, it seems that the different ways we experience the world would suggest, or give rise to, different patterns of language in each culture, while it would be equally true that the different patterns of language a culture has would give rise to different ways of experiencing the world around them. In this case language becomes an important factor as it has a great influence on a people's culture and our understanding of that culture and, ultimately, their philosophy.

In this connection we may safely say that when one learns a particular language different from one's own language, such exercise is likely to bring one closer to the culture and philosophy of that linguistic community. Thus, when African scholars learn foreign languages such as English and French, their understanding of these languages brings them close to the English and French cultures and philosophies. Therefore, it is to be expected that the understanding of a particular language is necessary for a thorough understanding of a people's culture and philosophy precisely because the knowledge of such a language induces reality in a way quite similar to that culture. This seems to suggest that the understanding of two or more languages is an inducement to the understanding of two or more cultures which, in effect, leads to the understanding of two or more kinds of belief systems and, finally, to two or more kinds of philosophy. We can assume that if this results in a conflict in cultures, beliefs, and perceptual references about the world, the problem of reconciling the conflict may stare us in the face. But if this situation arises, the question of which of the conflicting philosophical systems to adopt would depend largely on such things as emotional attachment to a particular language and culture, philosophical temperament, ethical, political, and scientific considerations, and even national pride.[13]

At present, none of the African languages is satisfactory enough to be adopted as a continental language, rich enough for analytic philosophy and science. Most of the advanced countries of the world have succeeded in spreading their ideas and cultures, especially by means of their philosophy, science, and religion, to other parts of the world through their well-developed languages. Curiously there exists a mutual assimilation of the so-called civilized languages of Europe into one another. This mutual assimilation has facilitated the easy translation of one of these languages into the other. Consequently, there is likely to exist a sort of cultural affinity which in effect is likely to result in a somewhat similar perception of reality among the users of these languages. As assimilation of one language into another depends on a parallel development of logical grammar, while a parallel development of logical grammar is seen by some philosophers as a result of the similarity of ways of life, general outlook, religions, social beliefs, and the scientific and nonscientific activities of different linguistic communities.[14] Perhaps in the face of the problem of a mutual assimilation of one African language to another there is need to develop an African continental language such as Professor Wole Soyinka suggested at the African Festival of Arts and Culture (FESTAC) held in Lagos, Nigeria, in 1977.[15]

For many Africans it is easier to learn English or French than a second African language because the former are well developed. But then it is through the understanding of these languages that many educated Africans today have become immersed in British and French philosophy and cultures long after the termination of colonial rule. From this it appears that the best way of propagating a people's philosophy and culture is through

18

their language. This is why today there is the juxtaposition of two cultures side by side--African culture and English or French culture--with the latter having an edge through the adoption of English or French as the official language in most of the African countries.[16]

Although an African mathematician or physicist understands the work of his European or American counterpart, he does so only because he is able to read and understand such work, not in his native language, but in a foreign, well-developed, scientific language. In spite of the fact that natural language is too poor to express many of the things scientists would want to express about the world, science has nonetheless advanced through well-developed languages. Hence, the poverty of African languages has led to the poverty of scientific ideas and meaningful contributions to the development of philosophy, science, and technology. One of the greatest problems of African thinkers today is to find the words in their respective languages or dialects that would catch precisely the meaning and reference of foreign words such as the scientific terminologies: physics, atoms, electrons, molecules, force and field, electromagnetism, thermodynamics, and even mathematics. In Yoruba the word mathematics refers to isiro, and this simply means arithmetic: addition, substraction, multiplication, and division. But then arithmetic is only a branch of mathematics. Although it is very easy for the British and German mathematicians to use the words "derivative" and "ableitung" to mean the same thing because their definitions are exactly the same in English and German, it is very difficult to find a word in an African language whose definition would catch precisely the same meaning as this mathematical concept. This, then, is an important problem to be solved if Africans are to make their own contributions to scientific development.

At present the troubles surrounding many discussions on African philosophy, socialism, politics, and traditional medicine arise because many of the concepts used are foreign concepts. The words "philosophy," "socialism," "politics," and "medicine" are themselves not African words, and discussions on each of these are usually not done in African languages. So if one enters into a debate on what is meant by African philosophy or African socialism, one must look at the words 'philosophy' and 'socialism' from the point of view of its meaning in English and what that would mean precisely in an African language. On this ground alone, all talks about African philosophy or African socialism is bound to result in conflict of meaning, cultures, and perception of reality. If there are accurate translations from an African language to English or French and vice versa, perhaps it would be clear what precisely is meant by these concepts in both languages and cultures. But as it is, most of African thought in philosophy, socialism, politics, and traditional medicine are written in foreign languages. This has led to the situation where serious efforts are being made by African intellectuals to make their thoughts fit into the thoughts and pictures of reality of the owners of these foreign languages.

In African philosophy, the influence of foreign language and culture persists, as could be seen in the two kinds of philosophical schools of thought among contemporary African philosophers; the Anglophone and Francophone schools. Finally, it is from this point of view of language that African philosophers are able to understand as well as teach British empiricism and continental rationalism, with all the conceptions of reality depicted by British and continental European languages about the world, society, and culture, religion, politics, science, and ethics. What is said about African philosophers is also true of African medical doctors. English and French have so gained the upper hand in their minds that many Africans have almost become foreigners to their own cultures. Consequently, the kind of power that derives from a belief in one's culture and system of thought, such as have helped Japan, China, and India to develop on their own, is almost nonexistent among African thinkers. In its place is a vaguely diluted culture, the product of which is an almost incurable self-defeatist mentality. Virtually all the Western-trained contemporary African philosophers in Africa today think, teach, and write philosophy in either English or French.

The same is true of people in the medical profession. And when it comes to African philosophy and traditional medicine, the official language of these subjects is not an African language but English or French. Under this pitiable condition it is quite tempting to forget or even disown one's culture as well as the belief and thought systems which are products of that culture. As a result it very often happens that some African philosophers and medical doctors think, teach, and write African philosophy and traditional medicine as if they disowned their own original culture to the extent in which they appear no longer capable of recognizing the potential contributions which their own cultures can make to philosophical and medical knowledge. It is a good thing to be critical of one's culture and belief. But when criticism is seen as tantamount to a wholesale rejection we should suspect that something is wrong, especially as we have seen that some African intellectuals have allowed foreign languages and cultures to seize the better part of their mental faculties. Under this condition any African thought must have a Western recognition or be seen as a Euro-African thought for it to be accorded a modicum of intellectual respectability. It is precisely for this reason that any field of human knowledge described as "African" is not seen as acceptable to Africans themselves unless it has the official stamp of a recognized foreign culture. I take this to be the greatest problem of African intellectuals today. It explains the essential difference between an African mind and a Japanese, Chinese, or Indian mind, and the reason why, in spite of its well-known intellectuals, Africa as a whole has not developed like these three nations. That is why Africa remains the only sleeping giant that has still refused to wake up!

As we said earlier, a people's culture and philosophy depend on the circumstances of their lives. The circumstances can be geographical, like having to survive in a harshly cold continent, or always basking in the sunshine without the need to think about the dangers of extreme cold weather. It could be war, religion, or certain facts of history. But, in every respect, circumstances do bring people together, especially when they share a common experience in space and time, to exhibit their cultural belief and philosophical temperament, both of which would normally force them to find common solutions to the common problems that threaten the survival of their society. The history of philosophy from the ancient through the medieval, enlightenment, and Victorian period to the present day highly technological societies can be seen as a history of changes in cultural beliefs, world views, attitudes to life, nature, and environment. All these are due to changes in philosophical temperament at certain points in time. Where such changes do not occur cultures, world views, and all the beliefs associated with them may be regarded as static. Those who have argued that African cultural beliefs and world views are static probably did so from this point of view--a view founded on an alleged non-empirical and noncritical nature of African philosophy.[17]

From the foregoing it would appear that our beliefs which are closely associated with our languages and cultures govern our practical lives and principles. These in turn are governed by the kind of world we think we are living in and dealing with. And what we deal with must, from all practical consideration, influence our judgment about reality. Such judgment about reality in turn creates our philosophy. It is from this point of view that we do have our theories of knowledge. The question of how to establish these beliefs will in effect be associated with the ways and methods of our search for the truth, depending on our philosophical temperaments. Which way we adopt for our search for the truth is itself subject to our way of life, our attitude, our perception of reality, and what we consider to be our relation to the external world, including human beings. In all this, culture plays an important role. Granted this essential relationship between philosophy and culture, the reader of philosophical literature must be impressed by the different and opposing philosophical systems in different parts of the world. Chinese cultures are different from British cultures in much the same way their philosophies differ. The same is true of French, American, Indian, and African cultures and their philosophies. I conclude, therefore, that such differences in cultures do provide at least a marginal academic respectability for different philosophical systems.

Chapter 3

THE QUESTION OF AFRICAN PHILOSOPHY

On Philosophy in General

Although a definition of philosophy is still being awaited, philosophy is seen as an attempt to arrive at reasoned answers to important questions.[1] This is only one definition of philosophy, but with it an important point could be made with respect to the questions and answers that characterize philosophy. Philosophical questions are not questions to which yes or no answers can be given readily. In this sense there can be no decisive or definitive answer to any philosophical question whether in metaphysics, epistemology, logic, mathematics, science, ethics, aesthetics, or religion.

According to Bertrand Russell,

> the conception of a life and the world which we call "philosophical" are a product of two factors: one, inherited religions and ethical conceptions; the other, the sort of investigation which may be called "scientific" using this word in a broader sense.[2]

He then noticed that "individual philosophers have differed widely in regard to the proportions in which these two factors entered into their systems," but maintained that "it [was] the presence of both, in some degree, that characterize[d] philosophy."[3] Yet, Russell regarded philosophy as something intermediate between theology and science: "Like theology, it is concerned with speculation, but like science, it appeals to human reason rather than authority."[4] The interesting conclusion from Russell's equivocation is his difficulty in giving a precise definition of philosophy. He in fact ended up by saying that philosophy is neither science nor theology for, as he said: "But between theology and science there is a no-man's land, exposed to attack from both sides; this no-man's land is philosophy."[5] To show that philosophy is not science, Russell emphasized the point that "almost all the questions of most interest to speculative minds are such as science cannot answer," for "science tells us what we can know, but what we can know is little, and if we forget how much we cannot know, we become insensitive to many things of very great importance."[6] Although he came from the

23

tradition of analytic philosophy, I believe that in saying this much Russell was aware of the importance of non-scientific questions in the minds of thinkers of all nations which philosophy tries to answer.

Let us now think about these important philosophical questions which science cannot answer, but which are of the utmost interest and importance to us. Russell provides some good examples:

> Is the world divided into mind and matter, and if so, what is mind and what is matter? Is mind subject to matter or is it possessed of independent powers? Has the universe any unity or purpose? Is it evolving toward some goal? Are there really laws of nature or do we believe in them because of innate love or order? Is there a way of living that is noble, and another that is base, or are all ways of living merely futile? If there is a way of living that is noble, in what does it consist, and how shall we achieve it? Is there such a thing as wisdom, or is what seems such merely the ultimate refinement of folly?[7]

Yet the list of important philosophical questions could go on to include others: why is there something rather than nothing, is reality one or many, what is an experience of God, what is His nature, is the world finite or infinite, is death the final end of life or is there life after death? From the above it appears that philosophical questions are mostly formulated in such ways that no indisputably correct answer can be given to them. "To such questions," Russell admits, "no answer can be found in the laboratory." Yet, "the studying of these questions, if not the answering of them, is the business of philosophy."[8] Even when answers are attempted, the best of them only succeed in raising further questions to which philosophers provide attempted answers or solutions.

Unlike science and mathematics, no problem ever gets solved in philosophy. We can therefore say, along with Kant, that in philosophy reason asks questions which reason cannot answer. It is precisely because of this that Professor Joad describes a philosopher as an impossible possessor of impossible knowledge.[9] Thus philosophers are of different temperaments, and their approach to philosophical issues are usually not the same. What appears to be agreeable to many philosophers is critical discussion of philosophical issues backed by reasoned arguments. But there is no agreement on the subject matter of philosophy. Hence, there are philosophers who either try to make sense out of nonsense or nonsense out of sense by reasoned arguments. When a philosopher who sits on his chair eight hours a day goes on to argue that tables and chairs may not exist after all, it looks to me that such a philosopher is trying to make nonsense out of sense. He could be replied to in the manner of Dr. Johnson by asking him to kick the chair with his bare foot to see whether there is a chair or not. The philosopher may also try to make nonsense out of sense by arguing that moral statements or statements about God, destiny, soul, or after-life have no meaning because they are neither true nor false. In this case, the propositions "killing is wrong" or "God is merciful" are nonsensical because they express meaningless statements, many of which are seen as commands, like "go away" or expressions of emotions like "Yeh!" or "Wow!" More interesting is that the philosophy of making sense out of nonsense and making nonsense out of sense have protagonists and antagonists.[10]

The general situation in philosophy now is that we probably do not know what to consider as a serious philosophical topic. Philosophers of the analytic bent spend such a great deal of time writing on such obscure topics as "meaninglessness," "nonsense," "conceptual absurdity," "as if," and so on that one begins to wonder whether philosophy is an analysis of language or a mere play on words. In whatever way we look at it, one cannot but agree with Waismann who found something "deeply exciting" about philosophy. "It is not a matter of 'clarifying thoughts,' or of 'the correct use of language' nor of any other of these damned things." What is it? "Philosophy is many things and there is no formula to cover them all."[11] This is probably why we can safely say, in the manner of Keynesian economists, that if philosophers were laid end to end, they probably would not reach a conclusion.

From the above we like to establish two things. First that philosophy as first described above is not empirical science or analytic philosophy pure and simple. As making nonsense out of sense, philosophy is either logical or linguistic analysis, a weapon for an attempted elimination of metaphysics which, ironically, is one of the most important branches of philosophy. In fact, from the philosophical point of view we can say that the program of the linguistic or the analytic philosopher is nothing but a tragic paradox of the philosophical enterprise. This is so because it looks as if the thing that gives philosophy its existence is metaphysics, the first philosophy.

As we can see, most of the questions raised above are metaphysical questions which by their philosophical natures are unanswerable. Yet people continue to ask metaphysical questions because the issues raised in them are perhaps the issues that touch on human curiosity to get answers to questions that reason cannot answer. The importance of metaphysics is shown by Fichte who declared that "the primary task of philosophy is to answer the question: What is man's destiny, his purpose, in the universe?"[12] But analytically-minded philosophers would contend that the majority of the questions that are in the minds of people turn out to be meaningless questions when considered from the point of view of logical or linguistic analysis. These are usually philosophers with different temperaments, or scientifically-minded Western philosophers. To them ethics, metaphysics, religions, and existentialism would be seen as irrelevant to proper philosophical discourse, that is to their own brand of philosophy--analytic philosophy which some philosophers have said is not philosophy but logical analysis of science.[13] In fact, analytic philosophy could be seen as a form of dogmatic philosophy. Philosophy, according to Collingwood, comprises a scale of forms each of which is a scale of dogmatic philosophy.[14] "It is by the road of dogmaticism that we can set forth upon the road to philosophy."[15] Even the ultimate aim of arriving at a complete theory of knowledge or explanation of the world is the ideal limit of dogmatic philosophies. Thus, "every person who is actually absorbed in any given form of experience is by this very absorption committed to the opinion that no other form is valid, that his form is the only one adequate to the comprehension of reality."[16] This is true of different philosophical schools: idealism, empiricism, rationalism, logical positivism, skepticism, pragmaticism, existentialism, mysticism, spiritualism, Buddhism, Marxism, Socialism, and Democracy.

The most essential feature of philosophy, according to Waismann, is _vision_. "From Plato to Moore and Wittgenstein, every great philosopher was led by a sense of vision."[17] Plato's philosophy, which is still very much with us today, attests to the truth of Waismann's position. Plato, in his Theory of Ideas, sees the exactness of mathematics as an abstract logical exactness which is lost as soon as mathematical reasoning is applied to the actual world. For this reason, Plato and many of his followers thought that, since mathematics is in some sense true, there must be an ideal (transcendental) world where everything happens as it does in the text books of logic and mathematics. The implication of Plato's theory is that when the philosopher, who is probably the only one who can get to Plato's heaven (Plato's ideal world), gets there, he is sure to be entertained with choice sights of everything he had missed in this world, like the perfectly straight lines, exact circles, triangles and, of course, perfect justice.[18] Plato's philosophy has all the profundity you can think of, but it is not open to refutation. Russell, one of the admirers of Plato, calls this philosophy vision, in much the same way Waismann does.

African traditional thought also holds positions quite similar to Plato's and some of those mentioned above. While some Africans look up to experience as the main source of knowing, others look up to something other than experience, for example insight, intuition, and the like. Yet others are imbued with downright skepticism. In African thought some ways of knowing, like experience and intuition, correspond to those in Western philosophy. Many Africans hold a view similar to Epicurus, a Western philosopher, in which case these thinkers are not likely to believe in life after death. By and large, these kinds of thinkers are those who would put greater trust on the evidence of their senses than on some transcendental entities. Some, on the other hand, do believe in a world of vision. This world, our

26

world some Yoruba say, is a market place. We would all go back to the other world--the ideal world--sooner or later. To these Africans the world, like Plato's empirical world, is a mere appearance, or, in Kant's terminology, a phenomenal, as opposed to the noumenal, world. This may explain fundamental relationship between an African and his belief in the after life, the reverence for the Deity and his usually generous attitude to people. Surely some look at the world as a place to enjoy while believing that after death they return to dust or atoms. These people do not believe that the soul is separable from the body, if there is a soul at all. Others conceive of the world as an appearance, a mere temporary domicile where one has to undergo or tolerate all the difficulties which are, to them, merely temporary.

For this kind of African idealist philosophy, we find an analogous case to Plato's. Like Plato, they think that since joy and sorrow are in some sense real in the world, there must be an ideal world of joy, a sort of paradise where everything happens as it ought to happen in the sensible world. The African idealist philosopher, when he gets to the other world where, presumably, only idealists (whether Greek, British, German, African, or American) go, will be rewarded by the existence of every joyful thing he has missed on earth and whatever else is necessary to perfect his bliss.[19] Now, if this kind of African traditional thought is unphilosophical, so must be Plato's. And if Plato's thought is seen as profoundly philosophical, surely the African thought is profoundly philosophical. Both exhibit some sort of visionary power which, in the words of Waismann, is the one single aspect and the most essential feature of philosophy. Our example from both Russell and Waismann on Plato seems to show that there is more to philosophy than empiricism or logical analysis of language.

Any discussion on the question whether or not there is an African philosophy must therefore take into consideration the nature of philosophy and the kinds of questions it raises. Surely, if philosophy is to be identified with empirical science or with logical or linguistic analysis, then only two branches of philosophy would stand recognition, viz logic and epistemology, both of which are philosophical tools for the analytic philosopher of science. At best, for other branches of philosophy to be philosophy properly so-called, they must employ these analytic tools. But then, what happens to metaphysics and ethics, both of which are also recognized as core areas in philosophy? The questions to be asked now are the following: Is a philosophy not good philosophy unless it is, or at least consists of, logic and epistemology? Are metaphysical and ethical issues and discussions not philosophical issues and discussions? Finally, must a world view be scientific in order to be a philosophical world view?

A world view, whatever its form, substantiates principles of different sorts: ethical, religious, philosophical, scientific, sociological, political. Hence a world view is a wider concept than philosophy. There are bound to be different world views, and philosophy is one of them. And if philosophy is a world view, it is certainly a world view, sui generis, of its own peculiar kind, that is a philosophical world view.[20] Now the question as to whether an African world view is philosophical or not depends on the kind of problems to which philosophy directs itself. Admittedly, if a philosophical world view is a scientific world view then we may doubt whether the African traditional thought or world view is philosophical. But as Oizerman puts it,

> The philosophical world view is above all the posing of questions. These questions arise not only from scientific researches but also from individual and socio-historical experience. . . . They may be called the main questions because, in posing these questions, philosophy enters upon a discussion that is important for all mankind. Such, for example, are the famous questions, the solution of which, according to Kant, constitutes the true vocation of philosophy: What can I know? What must I do? For what may I hope?[21]

Yet, these questions, according to Oizerman, do not exhaust the content of a philosophical world view. In answering these questions, people do have certain beliefs which they either hold a priori or, as Russell indicated, by the circumstances of peoples' lives.[22] Surely many philosophers held certain a priori beliefs about the external world until David Hume came to shatter such beliefs by his criticism of induction and experience.[23] But even in the face of Hume's powerful criticism many philosophers never gave up their belief in induction and their conception of the world. Even in political philosophy every theorist wants his own world view to reign without competition. Marx's belief in a Marxist world view has never, and will never, be given up no matter how the Western world may want their own over-arching capitalist world view to reign without competition. There is no a priori method of showing which philosophical world view is better than the other.

If we argue from the historical point of view that African philosophy has no tradition because there are no known figures whose philosophy could be studied as there are in the Western philosophical tradition, we may be misled into believing that Africa has no philosophical tradition. When one talks about British philosophy, the names of great figures like Bacon, Hobbes, Locke, Berkeley, Hume, Russell, and Ayer readily come to mind. For continental philosophy we know of great figures like Descartes, Leibniz, Spinoza, Kant, and Hegel. For American pragmaticism we can cite Charles Peirce, John Dewey, and William James, among the known great figures. The question that could be raised in connection with African philosophy is the same: who are the great figures to whom one might refer? But it would be a hasty judgment to conclude from the absence of great figures in African philosophy that African philosophy has no tradition or that there is nothing we can call African philosophy. For our purpose we would like to argue first that there is an African philosophical tradition and that the absence of great figures in African philosophy does not entail the absence or non-existence of African philosophy.

However, I would like the reader to note that nothing in our paper should be read as a claim that African philosophy is as systematic as Western philosophy; nor can we find in it even an inkling of analytic rigor that one finds in the twentieth century philosophy. Perhaps the latter is not an essential ingredient of all philosophies, as we shall see later on. But it is sufficient to admit that saying anything like the above would be too wild a claim to make for the emerging field of African philosophy, judging, at least, from the long tradition of Western philosophy itself. Against that tradition there seems to be no African equivalent. All the great figures that have been studied in the history of Western philosophy, either from the point of view of single nations or many nations taken together as a continent, the influence those figures have on other philosophers, and the relevance of their philosophical thoughts to contemporary issues, seem to suggest that a history of philosophy is perhaps a history of individual philosophers who had done and written philosophy over a period of time. For this reason alone it could be reasonably argued that, unless we can identify great figures with what we call African philosophy, there can be no history of African philosophy. This point would be well taken. But it can also be argued that a history of African philosophy can be seen from whatever it reveals about the foundations of African philosophy itself.

As it is, it is very tempting to base the argument against African philosophy on the absence of individuals or figures whose written work might be described as philosophy. But not all philosophical thoughts are written. Socrates, Buddha, and Confucius never wrote. Yet their thoughts became known to us through the writings of others. The recognition of the thoughts of these ancient thinkers as philosophical, be they on examinations of concepts or views drawn from religion, art, politics, morality, body and soul, life and death, and any other subjects of philosophical interest, seems to be a good example of discovery of the philosophical thoughts of earlier thinkers, although it might be argued that these thinkers were not unknown. But let us agree that what we are dealing with in the circumstances is thought, whether as anonymous or identified with a particular thinker. What is obvious is that there cannot be a bad or good thought without a thinker, and this will be true whether the thinker is known or unknown. Socrates was not an unknown Greek philosopher because Plato identified the source of his own philosophy with the earlier thoughts of Socrates. Nowadays it is difficult to make a distinction between Socratic and Platoric thoughts. Surely Plato gave Socratic thoughts some Platonic twist, to the extent in which Plato's dialogues might be said to have represented his own thoughts, though a refinement of earlier thoughts of Socrates, although the accuracy of Plato's recording of Socrates thoughts cannot be guaranteed. Here it is the thought of Socrates, either as original thought or a critical appraisal of earlier thoughts to which he also could have given his own original twist, that provided the material for Plato's philosophical inquiry. This original thought need not be as philosophically sound or rigorous as the further thought to which it gives rise. The original thought could even be made philosophically more complicated by the introduction of other thoughts from other geographical areas, such as we know of the Egyptian influence on Plato's doctrine of the immorality of the soul.

Perhaps some African figures of the Socratic stamp might have gotten their philosophical thoughts permanently recorded were it not the case that, in the absence of writing facilities, Africans kept their thoughts and those of others, including names and events, in their memories. These were usually passed down to other generations who learned them by rote. Under this circumstance, thoughts and events may fairly accurately pass down from one generation to another while the identities of persons are forgotten or regarded as unimportant. This does not prevent other people from using materials from unknown thinkers as a basis for philosophical inquiry, if the materials contain some philosophically interesting issues, as Plato did to the earlier thought of Socrates. The only difference is that the African cannot identify the source of his thought. Therefore it is this lack of identity of earlier African thinkers that is responsible for most of the problems the contemporary African philosophers are facing today.

In his paper "The African Philosophical Tradition" Lancinay Keita suggests a division of the African philosophical tradition into three phases:

1. classical African thought concerning itself with the thought systems of ancient Egypt and their influence on the Hellenic world and later on European Renaissance;
2. medieval African thought, focusing on the African interpretation of Islamic thought in the medieval African States of Ghana, Mali, and Songhay;
3. modern African philosophy, which Keita curiously sees as the least developed of the three, "since philosophical traditions have become somewhat distorted as a result of the colonial experience," with the best works leaning towards politics and literature.[24]

Keita's first two phases correspond to mine, with little or no evidence of great figures comparable to Western philosophy, ancient or medieval. While I recognize only St. Augustine as the great person during my first phase, I argue that there might be other people not only in the northern part of Africa but in Africa south of the Sahara. The works of anyone cited would of course depend on the philosophical merits of such works which would give the authors the qualifications of leaders in African philosophy. In the absence of such leaders, the period will be regarded as uneventful even though some Africans who lived at this period might have practiced philosophy or indeed influenced the ancient Greek philosophers. This, in short, is what characterizes my first phase which spreads over a long period of time. Keita's third phase corresponds to mine, although mine is not less developed than the previous phases. Although contemporary African philosophers criticize their known predecessors, mainly anthropologists and missionaries turned philosophers, for the first time they have entered into a rigorous debate as to the existence of African philosophy.

In what follows I shall discuss philosophy in Africa from a historical point of view based upon insufficient sources, and suggest three phases: (1) unwritten philosophy and unknown philosophers, (2) re-orientation in philosophy, and (3) critical re-orientation in philosophy. The first phase was very uneventful, although it could have provided some materials as food for subsequent thought. The unwritten nature of African traditional philosophy and lack of identity of individual thinkers seemed to have given room for all sorts of manipulation of African thought by some colonial scholars who, from their writings, are known as ethnophilosophers. They belong to the second phase of our history as suggested above. To the third phase, critical re-orientation in philosophy, belong contemporary African philosophers who had undergone a period of philosophical training in the Western world, mainly in the English and French-speaking countries. This new group of philosophers look at philosophy in quite a different way from their colonial predecessors. By looking at philosophy with purely analytic eyes they not only dismiss the previous conception of African philosophy as false, they end up almost by making the concept of African philosophy either problematic or impossible. I say "almost" because they probably believe that the debate on whether or not there is African philosophy is itself to be known as African philosophy, the very subject of the debate or controversy. Were it not the case, such a debate ought not to have been written and published as African philosophy. Be this as it may, this exercise by itself could be seen as the beginning of an attempt to give a definite meaning to African philosophy, if one exists, and the need to write and teach it as an academic discipline in African departments of philosophy.

The teaching of philosophy in Africa today is more of the teaching of Western philosophy than African philosophy. People tend to be more productive in the writing of African philosophy as controversy or debate on whether or not African philosophy exists. But then some of those who debate the existence of African philosophy consciously or unconsciously teach it as a course on controversy. This is also the case for African philosophers who merely talk or write about its possibility, but teach instead African traditional thought in their departments of philosophy. My own contention is that a philosophy should be done in a positive way if it exists, and not endlessly debated upon if it does not. For instance, in Nigeria where there exists the largest number of universities as well as the largest number of philosophy departments in the whole of Africa, African philosophy is not only taught, but is a compulsory course.[25] When some African departments of philosophy still retain the title "African traditional thought," and philosophers only talk about African philosophy in their debates, their Nigerian counterparts, together with some expatriate philosophers, have gone ahead by teaching, researching, and writing on African philosophy. The exercise which began around 1978 has produced graduates and postgraduates in African philosophy, at least from the university of Ife in Nigeria. How, then, did philosophy begin in Africa? What are its stages of development and what is the general situation of philosophy in Africa today? We shall attempt to answer these questions by looking into the three phases of development as suggested above, from the vantage point of the doing, writing, and teaching of philosophy.

A Short History of African Philosophy

In Africa philosophy is an emerging field, relatively new but making its impact felt. One cannot discuss the activities mentioned above without a recourse to their past, their influence on the present, and possibly their effect on the future.

The activity known as doing philosophy is not necessarily writing philosophy. Socrates, Buddha, and Confucius, who are usually referred to as the Paradigmatic Individuals, did philosophy and probably taught it, although not in the formal sense of nowadays, without writing it.[26] Therefore, while the exercises known as doing and teaching philosophy are not necessarily writing philosophy, writing philosophy is doing philosophy in a strong sense, and also teaching it in an important sense, i.e., in the sense in which philosophical works are used for teaching philosophy through reading of them as philosophical textbooks. From this analysis it appears that the most important of the three philosophical activities is writing philosophy while doing or teaching philosophy may not involve writing it. The problem of the history of African philosophy can be seen from this point of view. Although some Africans might have done and even taught philosophy of some kind, the lack of written philosophy is likely to make it difficult for people to understand what is meant by African philosophy or its history.

32

But the point still remains that people can do philosophy without writing it since, as we also have said, doing philosophy is not necessarily writing philosophy. However, if not all history is recorded history, and not all thoughts are written by the original thinkers, we can explain what we mean by a history of African philosophy. It does not matter if we call it a history of philosophy in Africa or a history of African philosophy. I do anticipate that in the future others may come up with a historical view of African philosophy slightly different from mine. Hence I call mine a history, rather than the history, of African philosophy as seen from the point of view of its historical phases.

If doing philosophy is not necessarily writing philosophy, then the absence of written works by great African figures on African philosophy that are comparable to the known works of great figures in the history of Western philosophy is no proof that African philosophy is non-existent. For all we known, African thinkers might have been doing philosophy without knowing that they were doing it. If somebody now comes into contact with those who were in contact with other people who had been in contact with the unwritten thoughts of ancestors who speculated about the world around them, about the place of human beings in the universe, persons as body and soul, or body, soul, and inner-head, people and their destiny on earth, truth, knowledge as personal or impersonal, principles of conduct and human association, and put into white and black the philosophical contents and issues raised in these hitherto unwritten thoughts, such a person could be said to be writing the philosophical ideas found or discovered from the thoughts of his ancestors or ancient thinkers. Very often such materials provide us with the basis for doing and writing philosophy, even though in the end we may improve on this philosophy by injecting our own philosophy into it, such as Plato probably did to Socrates. But the point to be remembered here is, as Russell said, you have to be to some degree a philosopher before you can discover the philosophy of a person or a people.27 From this point of view we see a history of African philosophy in three phases.

(i)   First Phase: Unwritten Philosophy and Unknown Philosophers

Because we have no evidence as to when unknown philosophy started, we can only speculate that the exercise of doing but not writing philosophy in Africa had always been with the African people. From the unwritten nature of such a philosophy many important thoughts might have been lost or forgotten, especially as the means of passing thoughts from one generation to another was learning by rote. This leaves us with oral evidence as the best available means of getting into the ideas of these ancient thinkers. The danger of doing philosophy without writing it is that the idea dies with its owner unless it is recorded or learned by rote. A similar danger exists for learners by rote, as the ideas which they themselves had stored in their memories die with them. In spite of these difficulties, there still are some ideas which have survived and were fortunate enough to be alive at the time when Western education came to Africa. But it was not until the first half of the twentieth century that the thoughts of African peoples began to find their ways into print. It is for this reason that the first stage of African philosophy could be said to be very uneventful, more so as the identities of individual thinkers were unknown. Therefore, when these thoughts passed down to Western anthropologists and missionaries, they were treated as the collective thoughts of African peoples. This became known as ethnophilosophy.[28]

As I said earlier, there is no accurate record of the time when the exercise of doing philosophy by Africans actually began in any part of Africa south of the Sahara. There is, of course, substantial evidence that in North Africa philosophy was done and written between 570 B.C. and 430 A.D. Pythagoras, who had a tremendous influence on the development of Plato's doctrines of the immortality and transmigration of the soul, got the idea from his contact with Egypt.[29] Pythagoras may well have got the idea from oral evidence as passed down from generation to generation. Socrates also speculated about the human soul, but it was not until Plato that any substantive work was written on this philosophical doctrine. St. Augustine, an African born in Thagaste in North Africa around 354 A.D., was probably the first African who did and wrote philosophy. His philosophical writings showed that he believed that knowledge of any kind is a function of the soul or mind, and defined soul in the Platonic manner, as "a substance endowed with reason and fitted to rule the body." He agreed with Plato before him in his definition of man as "a rational soul using a mortal and material body," while he conceived of the relation between soul and body on the model of the ruler and the ruled, or the user and the tool.[30] What is not immediately clear is whether St. Augustine was ever influenced by Plato's work (if Plato's work had existed in writing in any form at that time), or whether he simply was working on and writing down the ideas he got from North Africa where he was born and lived, and from where Plato also got the same idea through Pythagoras.

Now, since there was a parallel idea and doctrine of the soul in other far away parts of Africa south of the Sahara, among the Yoruba-speaking people of southwestern Nigeria for instance and probably in many parts of black Africa, the question arises as to the historical origin of the doctrine of the human soul, that is, whether the idea came to North Africa from the southern part of Africa or vice versa, or whether the idea existed in northern and southern Africa quite independently. As I had treated this matter elsewhere, I do not intend to discuss it further in this work.[31] What is important to note, therefore, is that if the continent of Africa is taken as a whole, there is evidence of philosophical thoughts which were written down, dating to St. Augustine. On the other hand, if the phrase

"African philosophy" is to refer to black Africa alone, there is evidence that people had philosophical ideas similar to those in the Western philosophical tradition, although, unlike the situation in North Africa, there is no evidence that such philosophical ideas were written down. On the whole it appears that not much can be said about the exercise of writing philosophy at that period in Africa for, apart from St. Augustine, there were no known great figures. But this does not mean that there was no philosophy done that could later engage the attention of scholars. This period may therefore be appropriately called the period of unwritten philosophy and unknown philosophers in Africa.

(ii) Second Phase: Re-orientation in Philosophy and Colonial Ethnophilosophers

It is to be assumed that philosophy in a nontechnical sense had existed in African thought systems even before its discovery by colonial scholars, mainly European missionaries, ethnologists, and ethnographers on the one hand and indigenous Africans whose main interests were in studying African thought through their traditional religions and cultures on the other. The first "orientationists" in African philosophy belong to this group of colonial scholars who were themselves not philosophers and mistakenly identified African philosophy with traditional religions and cultures. If it is true, as Russell said, that you have to be in some important sense a philosopher in order to understand the philosophy of a people, we might argue here that all the fundamental misconceptions about African philosophy originated from the writings of European missionaries and anthropologists, and even some African writers who saw African philosophy only from the point of view of traditional religions and cultures. From their writings, no indications exist that these scholars were able to recognize the philosophical from the non-philosophical contents of African thought. As would be expected, they produced anthropological expositions of African traditional thought, studied from the point of view of what they called "primitive cultures," often described as prescientific and prelogical. Even those who managed to describe what they were looking for as philosophy were by no means technically competent to make a distinction between philosophical and nonphilosophical thought. To this group of colonial scholars whom I call orientationists in African philosophy belong Levy-Bruhl, the Reverend Father Placide Tempels, the Reverend Father Alexis Kagame, Janheinz Jahn, Robin Horton, and the post-colonial African theologian, John Mbiti.
Perhaps one of the most influential exponents of the idea of African philosophy as traditional religions and cultures, prescientific and prelogical, is the French ethnologist Levy-Bruhl (1857-1939). As a colonial scholar, Levy-Bruhl was sent to Africa by the Ethnological Society for Colonial Studies. His

work was peculiarly ethnocentric. Using Western standards to judge African systems of thought he concluded that African culture and thought systems were prescientific and prelogical, using the term prelogical to describe a kind of thought that is not free from inner self-contradiction. Although he later renounced his position in his posthumous Notebook on Primitive Mentality (1975), that renunciation was not as well known as his earlier classic, Primitive Mentality (1923, reprinted 1966), a book which might just as well have been entitled "Prelogical Mentality." Subsequent writers had followed his idea of African philosophy as ethnophilosophy, characterized as a collective thought or group mind of a people.

Reverend Father Tempels, the author of Bantu Philosophy, was a Belgian missionary in Africa who both rejoiced and lamented his so-called "discovery" of Bantu philosophy under the chapter entitled "Bantu Philosophy and our Mission to Civilize." Writing about the civilized and the uncivilized people, Tempels states as follows:

> If we are justified in the hope that we have plumbed the depths of the primitive soul in this treatment of Bantu philosophy, we shall be obliged to revise our fundamental ideas on the subject of "non-civilized" peoples: to correct our attitude in respect of them. The "discovery" of Bantu philosophy is so disconcerting a revelation that we are tempted at first sight to believe that we are looking at a mirage. . . . Ethnologists of the evolutionary school have already been perturbed by the "troublesome statements" of those who have revealed that it was amongst the most primitive peoples, those least civilized that the purest and most sublime idea of monotheistic God was to be found. Is not the discovery that there is such a thing as philosophy among the Bantu going to lead to other "troublesome statements" of the same kind? It would seem, in fact, that the erroneous deviations from and inadequate applications of Bantu philosophy noted in the body of this book are generally of recent date. Older Bantu thought, healthier and more certain, can still be discovered in its most exact form among the most conservative tribes.[32]

Tempels thought he was presenting a tribal philosophy, and in the end it was not clear whether his work was on philosophy or ethnology. Contemporary philosophers are apt to call it ethnophilosophy.

The other writers in this group did no better. Father Alexis Kagame, the author of Rwandan-Bantu Philosophy of Being had written his own book in a style resembling Bantu Philosophy, and about the same period as Tempels. Janheinz Jahn, a German

36

perspective of other peoples and cultures. This itself is an admission that a people's culture and language do influence their thought or philosophy. But here again, it is the business of philosophers to judge what thought of a people conforms to philosophical thinking. This is to say that you have to be a philosopher in order to understand the philosophy of a people or individual. "To understand an age or a nation," says Bertrand Russell, "we must understand its philosophy, and to understand its philosophy, we must ourselves be in some degree philosophers."[34]

From the above discussion it appears that if the re-orientationists achieved anything, it was the general confusion they created in people's minds concerning the nature and scope of African philosophy, although due credit must be given to them for re-awakening people's interest in the idea of African philosophy. But it was never really clear whether they were concerned with the real question of African philosophy or were just interested in discovering what they wanted to discover as African philosophy, namely, ethnophilosophy. Whatever they discovered as philosophy or nonphilosophy in African thought systems were so discovered, from the point of view of anthropological and ethnological studies and a preconceived idea of African traditional thought, as a product of uncivilized or primitive cultures from whose thoughts a philosophy must be discovered if only to show that Africans live in a different world of their own and are incapable of second order thought. Thus, any discovery of a philosophy that bears a resemblance to Western philosophy would be a troublesome discovery to anthropologists and ethnologists of the evolutionary school, a troublesome idea for colonial educationists and a disturbing event, particularly for those who were concerned with African education. This was so because it would mean that the mission of the colonial scholars to educate and to civilize, based on the preconceived idea that the whole of African minds was like a tabula rasa, would turn out to be a disappointing exercise.[35] As it turned out, the mission was not a disappointment, for where philosophy was discovered it was turned into ethnophilosophy, concocted, as it were, out of a philosophy of a group mind, a collective thought of a people.

It is obvious from the above that philosophers are in a better position professionally to judge which aspects of African thought is philosophical and which is not. Ethnographers can neither judge in advance or from field work that a people's thought is unphilosophical. Even in this case it is not clear whether they mean all thoughts or some thoughts, or the thoughts of some people or the thoughts of all people in a culture. But from their writings, anthropologists and ethnographers refer to the traditional thought of Africans as a collective thought, or group mind. This is the way the anthropologists use the word philosophy when they talk of Bantu philosophy, Rwanda-Bantu philosophy of Being, and Kalabari thought.[36] Even when a people's ways of thinking were recorded, they were presented merely as facts without any consideration for their philosophical implications. However, it is probably a philosopher himself who could best appreciate the philosophical implications of any system of thought. Thus, the overall situation of things during this period is that philosophy was alive but unwell in Africa.

(iii) Third Phase: Critical Re-orientation in Philosophy and the Contemporary African Philosophers

By critical re-orientation in philosophy I mean the period of reaction against the ethnophilosophers by the contemporary African philosophers who, by virtue of their formal training in Western philosophy, are in an advantageous position to judge what could be called African philosophy.[37] Perhaps the most notable contribution to philosophy in Africa today is the debate among the African contemporary philosophers on whether or not there is African philosophy.[38] However, I do not think that this contribution itself amounts to doing African philosophy, for all the talks and writings on the debate are exercises in Western philosophy.

If philosophy is to be done, to be written and taught in a modern African department of philosophy, as opposed to the past when it was taught in the departments of anthropology and religious studies, it must conform to an acceptable standard. The question is what standard, and whose standard, should be the acceptable one? The answer to this question is simple, judging from the academic backgrounds of African professional philosophers who belong to a relatively new academic field on the continent of Africa. Having undergone their rigorous philosophical training in the West, and having been acquainted with certain standards of doing, writing, and teaching philosophy, it seemed to them quite clear that philosophy must be done, if done at all, according to the Western tradition. For quite some time, therefore, notably between 1969 and 1977, the teaching of philosophy in some African universities consisted mainly of the traditional subfields of Western philosophy, namely, logic, epistemology, metaphysics, ethics, history of (Western) philosophy, and some special subfields such as philosophy of religion and social and political philosophy.

What we now call African Philosophy was taught as African Traditional Thought. During this period African philosophers of Western orientation (including the present writer) developed cold feet towards the idea of African Philosophy and retained the title of African Traditional Thought. Suddenly there appeared on the pages of journals a debate as to whether or not there was an African philosophy. Perhaps this debate might not have been necessary if the African philosophers themselves were sure that there was no African philosophy. Even if the debate was founded on some doubts about African philosophy, the debate ought not be an endless one. It might even be argued that perhaps a better way of finding out whether there is African philosophy is not to debate about it, but to critically examine the philosophical contents of African thought, some of which might have eluded the colonial scholars, either by virtue of their language handicap or simply their refusal to see them as philosophical. It was also quite possible that the anthropologists and ethnologists (all ethnophilosophers) were greatly handicapped, by reason of their training, in finding "appropriate canons of comparison" between two or more systems of thought, a comparison that could have led to a more positive approach to the issue of African philosophy.[39]

Instead of employing their philosophical training gained from abroad to discover whatever African thought was seen as similar in kind to those they are familiar with in Western philosophy, our African philosophers apply their Western philosophical training to arguments concerning the existence or non-existence of African philosophy, as if such arguments would bring it into existence or drive it out of existence as a discipline in African, or non-African, departments of philosophy. The present writer abstained from the debate because, even though his own previous approach to philosophy was conceptually analytic, he did not believe that the whole of philosophy could be defined from the analytic point of view. We cannot use the Western analytic philosophy which is only one of many philo-sophical traditions in the West, to dismiss other kinds of philosophy as non-existent. If this were to be done, a great deal of the contents of Western philosophy would be dismissed. Metaphysics and existentialism, for example, would be expunged from the discipline. It does not matter whether in these subjects many great figures could be named in the history of Western philosophy. What matters would be whether or not metaphysics and existentialism consist of verifiable statements, or are amenable to logical or linguistic analysis. Admittedly some aspects of African thought or philosophy do consist of unverifiable statements such as are found in metaphysics, existentialism, ethics, and aesthetics, the best critics of which belong to the now largely discredited school of logical positivism. But who says that unverifiable statements are non-philosophically meaningful statements other than philosophers of the analytic bent?

From the above point of view my own position is that we cannot, and must not, use the analytic rigor as the telescope and yardstick through which we look for and by which we measure the existence of African, Oriental, or any non-Western philosophy. Therefore, those who have persistently debated whether or not there is an African philosophy seemed to have taken the right step but in the wrong direction. It is a right step in that it led to the recognition of the possibility of African philosophy. The direction to which the debate is heading, however, is purely a negative one. It does not encourage any attempt at positive work on the philosophical contents of African thought. Engineered by this debate, I had examined some of the contents of African thought and found them quite similar to Western philosophy, such as the existence of God, concept of a person, and the problem of the mind-body dichotomy, immortality of the soul, human destiny, determinism, freedom, moral responsibility and punishment, human values, knowledge and probability, Ifa compared with Laplace's omniscient intelligence as a repository of knowledge, and the role of Ifa in medical knowledge.[40]

Judging, therefore, from the nature and scope of philosophy, although the definition of philosophy itself is beset with difficulties, I made suggestions about possible areas of African thought that are of great philosophical importance, even in the contemporary sense, in my criticism of Horton.[41] My paper rejected Horton's views, which were probably the most current at that time, that African traditional thought was unphilosophical because it did not meet the analytic ideal of Western philosophy. His conclusion was based on wrong premises. First was his misconception of philosophy as empirical science. Second was his idea that philosophy was nothing but logic and epistemology, and particularly logic. This post-Levy-Bruhlean idea seems to suggest that before the invention of logic by Aristotle there were probably no philosophers and philosophical systems, or that all it required for people to think logically, rationally, or philosophically was just this discovery of logic.[42] Important though it is, logic is just one of the traditional subfields of philosophy, and I believe that a great deal of work has been done in Western philosophy without the slightest indication that the writers' priorities were in logic. Yet it cannot be shown that there are any systems of human thought anywhere in the world in which the principles of logic (noncontradiction, identity, and excluded middle) are never employed in reasoning, either consciously or unconsciously. This is to say that these fundamental principles of logic need not be learned by, or taught to, people before they could reason logically.

Because of its assumed universal validity and applicability in all thought, logic, like mathematics, cannot be relativised. As there is no American or African mathematics, there is nothing to be called American or African logic, otherwise it would no longer be true that the principles of logic are assumed in all rational thought, a condition that applies without discrimination to all people. They, of course, would not be known as human beings but beasts to whom the above condition does not apply. It is one of the conditions by which we differentiate human beings from the lower animals, and that is why language is an important part of logic and reasoning. Where you have language there is logic, and where there is logic, there is language. By language here I do not mean language as it is used for expression and communication alone, otherwise there would be no difference between human beings and the lower animals, since both use language for _expression_ and _communication_. But more than these are the higher functions of human language; more especially the

41

descriptive and argumentative functions of language, as Popper
would say.[43] Therefore, if the principles of logic are universal
in all human beings capable of thinking in a human language,
coherent logical systems could be built in all human languages,
insofar as such languages not only perform the expressive and
communicative, but also the descriptive and argumentative,
functions. For instance, Modus Ponens, a valid principle of
inference, would be recognized as such in any African thought
system. The same could be said of other logical principles,
including the three fundamental laws of thought.
This would be true of all principles of valid inference.
To know this we just have to examine the languages of Africans,
and their use. In the Yoruba language, for instance, Modus
Ponens would be recognized as an impeccable argument if it were
given. For example

Ti o ba se wipe ojo ro, a je pe ile tutu. $p \longrightarrow q$
(If it rains, then the ground is wet.)
Looto ni ojo ro. p
(It rains, or It is true that it rains.)   _____
Nitorina, ile tutu. q
(Therefore, the ground is wet.) $\therefore$

"If . . . then" here is "Ti o ba se wipe . . . a je pe." Space
does not permit us to go further in this logical exercise. But
more of it, together with a discussion on logical connectives
such as 'and,' 'or,' and 'negation' are possible areas of future
discussions on African philosophy. We cannot, therefore, write
off in advance the possibility of a development of logical
systems in any language and thought of a people, even if no
logical systems have been built at present in that language.
From this point of view it could be argued that if there were
human beings on other planets and we were to understand their
language, and they understand ours, we could then communicate
beyond our own planets. If this happened, we would also understand the language, logic, and philosophy of other planets. As
long as the inhabitants of these planets are human beings, we
have no reason to deny them the higher functions of human
language.
On the other hand, if logic is not universal in thought, it
would then be perfectly conceivable that our own logic, which we
consider to be the paradigm of rational thought, could very well
be seen as irrational by the occupants of other planets. This
would, of course, mean that the universal character of logic is
suspect, and we may then look upon logic again as nothing but
relative to different thought systems. In this case it would
even be more difficult to write off a particular thought system
as illogical or irrational, and we would then be at the mercy of
an ingenious philosopher or omniscient logician to formulate for
us precisely what the criteria of rational thought should be
under this condition. It would no longer matter whether in one

42

thought system contradiction is not only possible but is actually allowed, as it is in fact allowed by Hegel, or the principle of excluded middle is seen to be unreal, for it could then mean that what we take to be the traditional principles of logic are, after all, relative to different peoples, cultures, and possibly planets.[44]

If logic could be relativized in this way, there seems to be no escape from an equally damaging conclusion against the arguments for the so-called prelogical mentality of the African thinkers. What the above arguments tend to show is that first, if we consider logic as universal in thought, we cannot deny it in any system of thought without denying also that it is indeed universal in thought. On the other hand, if logic could be relativized, we also cannot deny it in the thought system of others, for to do so would be tantamount to saying that logic, if relative, is not at all relative. And this is to bring us back again to our earlier conception, that logic is universal in thought. But logic is either universal in all thought or it is relative to different thought systems. So, in neither case can we deny logic in the thought systems of others.

But perhaps the anthropologists do not use the term "logic" in the way philosophers understand it. At this stage it would then be necessary for our anthropologists and ethnographers/ ethnophilosophers to formulate in precise terms what is meant by British, American, Indian, or African logical or nonlogical thought, the presence of absence of which could be related to the presence or absence of a rational language in any culture. Surely, from such an exercise we might come to understand the essential differences between a British logical thought and an African prelogical or nonlogical thought, or what our anthropologists mean by a "logical culture" as opposed to a "nonlogical culture."[45]

From the above, our position is that the case of pre-logicism made against African thought has not been proved to be true. Rather, our own argument seems to have shown that the reverse may very well be the case. If only our colonial anthropologists had not conducted their researches under a principle which I call "selective vision," that is seeing only what they wanted to see, and writing, with some exaggeration, on what they saw so as to fit them into their existing hypotheses or justify their preconceived ideas, some of them could at least have discovered a great deal of mathematical and logical thinking involved in some of the African games. It has been found, for instance, that the game of mankala, like ayo, wari, and solo, involves complicated mathematical and logical thought comparable only to the Chinese game of go. Some work is being done on these African games.[46]

We also would have expected the anthropologists and ethnographers to study carefully the philosophical issues in African thought as given earlier in this book, and see the appropriate comparisons that could be made between them and

Western philosophical thinking on the same issues. Perhaps it was not their business to look into the philosophical contents of African thought. From their professional points of view they might be pardoned. But philosophers, especially African philosophers, need not stop at merely rejecting the anthropologists' conception of African philosophy as ethnophilosophy. Their critical reorientation in philosophy, especially in African philosophy, should also not end with the current debate on the subject, as if to leave future African philosophers to untie the Gordian knot of mystery being created around the subject of their debate--a mystery which is not unravelled but actually aided by the seemingly endless, negative debate on the question of African philosophy. If we are not careful the exercise may end with contemporary African philosophers making no contributions to a positive development of African philosophy.

But when all is said it is my belief that the question of African philosophy can be settled by appropriate comparisons between some African thinking and Western philosophy by means of some African traditional thinking which address itself to much the same philosophical issues in Western philosophy. The question of how rigorously African traditional thought philosophically addresses itself to the same philosophical issues or problems in Western philosophy is lame, for it already presupposes that there is such a thing as the philosophical method. There is much of the traditional thought that could properly be called philosophy even in the modern contemporary sense of the word provided, of course, we do not make the mistake of identifying philosophy with either analytic philosophy or empirical science. It does not matter whether such thought is original with Africans or is borrowed from outside (as American pragmaticism was borrowed from the tradition of British empiricism) and modified by an original twist. Provided it touches upon ideas that are philosophically interesting, it is philosophical, and such thinking will not be called historical, geographical, or anthropological precisely because philosophy, however difficult it is to define, is neither history, geography, nor anthropology.

There are, then, at least some materials with which to do, write, and teach African philosophy, as I have earlier suggested. Many of such thoughts, which can also be found in the work of indigenous writers, belong to the discipline of philosophy although not of the analytical kind. If these thoughts have anything to do with philosophical thinking, they would be fit thoughts for philosophical treatment. To do this properly is the task for African philosophers themselves. But at present there are two categories of contemporary African philosophers: the negative or skeptical thinkers and the positive thinkers.

To the first category belong those philosophers who, by virtue of their training in Western philosophy, have become skeptical about the idea of African philosophy. While they rightly rejected the colonial anthropologists' conception of

African philosophy as group philosophy or ethnophilosophy, they have failed to discover, either by comparison or careful investigation, an authentic African philosophy. This, of course, judging from my argument concerning the language problem, is the permanent colonial legacy bestowed on African intellectuals who see every intellectual endeavor from the point of view of their training in the languages of Western cultures. These are the philosophers who have debated about the existence or non-existence of African philosophy for more than a decade, with some Western philosophers making their own contributions as moderators.

The main argument of this category of contemporary African philosophers rests on what they regard as the uncritical nature of African philosophy. They also assume, rightly or wrongly, that philosophy is simply a linguistic or conceptual analysis.[47] First of all no African philosopher who had his training in the West would dispute the fact that criticism and argumentation are essential tools in philosophy. But then even in the West, philosophy consists not only in criticism or arguing from premises to conclusion, but also in speculative thought. In some cases, criticism comes at the latter stage of enquiry. Some wise sayings, such as aphorisms of thinkers like Confucius, Seneca, Cicero, Socrates, Emerson, Jefferson, and others, have inspired philosophical ideas about politics, human relation and ways of life. Philosophical speculations can produce the seeds from which important philosophical discussions germinate. Philosophers everywhere must be capable of identifying issues that are philosophically relevant in any thought, be it Western, African, or Oriental. After all, Western philosophy in the twentieth century is different from that of the Medieval period, and I believe it will be different in a hundred years time, given people's changing perception of reality. I do not by any stretch of the imagination quarrel with the view that philosophy must be critical in its method. Philosophy has made progress through criticism. But criticism or skepticism can also turn out to be an obsession or a philosophical dogma, if critically considered. This point must be borne in mind.

It has been suggested that because African philosophy is not rigorous or exact, we might doubt its existence.[48] It is by no means clear what the exponents of this view mean by "rigorous" or "exact." For instance, how rigorous must a philosophy be before it is allowed to pass as philosophy? What, in fact, is an exact, as opposed to an inexact, philosophy? Is philosophy to be identified with logic or mathematics, the two rigorous disciplines usually referred to as exact sciences? Surely, behind the minds of the exponents of rigorous or exact philosophy is the kind of analytic rigor found in analytic philosophy and the philosophy of science or, if you like, the analytic philosophy of science. Judging from the history of philosophy and different philosophical movements, this is an extremely narrow, highly specialized, view about philosophy. Original philosophical

thoughts are not logical or critical analysis, or even argumentation. They could be a form of speculation about the origin of
the world and man's place in it, or the raising and answering of
such questions as suggested earlier. These are the kinds of
original philosophical thoughts that paved the way for critical
analyses or philosophical argumentations by professional
philosophers. A rigorous or analytic philosophy means no more
than a critical or logical analysis of the original thoughts and
concepts of others, whether ancient, traditional, or modern. As
we pointed out before, it may be difficult to say precisely where
the use of the word "traditional" stops and the "modern" begins
in the qualification of philosophical thinking. Surely much of
so-called traditional thinking has been given modern interpretations, and some could be found useful in solving modern problems.
Therefore, to dismiss traditional thinking as unphilosophical
simply because it sometimes does not consist of argumentation,
logical or critical analysis seems to me a bad judgment. The
philosopher would at least examine carefully this thinking, find
out which aspects of it is philosophically relevant and go ahead
to subject it to philosophical analyses. His analyses might end
up either in the improvement of this thinking, or its abandonment. In either case, the exercise, which can hardly be regarded
as final, would lead to further philosophical discussions on the
so-called traditional thinking.

Perhaps it should be emphasized that some philosophers have
even argued against analytic philosophy and concluded that it is
not philosophy, and its adherents are not philosophers but
logical analysts of science.[49] Although the constructive work
they have done is of great value, especially in the philosophy
of science, Professor Chatalian has argued that their work should
not be confused with the work of philosophers who are conceived
as ultimately struggling to work out a philosophy for humanity,
give an insight into a better understanding of the meaning of
life; a philosophy of civilization; a philosophy as "the guide of
life" for humanity as a whole.[50]

Before the end of this century analytic philosophy would
have reached the limit of conceptual rigor, and a new revolution
in philosophy taken place. Such a philosophy, which would take
humanity and the improvement of mankind as its central focus,
would not be insensitive to the general conditions of people. It
would appeal to people everywhere because of its involvement in
human values (traditional and modern), affairs and conditions of
people in a rapidly changing world, human goals in life and all
other issues that are of great interest and concern to humanity
as a whole. Such a revolution in philosophy which I believe is
long overdue, would take place with some positive contributions
from the non-analytic philosophies of other nations and conti-
nents, including Africa. The aftermath of this revolution would
not only lead to a down-to-earth conception and enrichment of
philosophy, but would also guarantee a better future for
philosophy and philosophers, from the point of view of their

relevance to society, university departments, and the need for continuous financial support and expansion.[51] This awareness, I believe, is perhaps sufficient to force a definite change of approach to philosophy and our conception of it as an unconditionally viable discipline. The change, as I conceive it, is potentially a revolutionary one which, from its general appeal, may turn out to be the most remarkable turning point in the history of philosophy. This must not be taken as an indictment against analytic philosophy, but rather as an indictment against any conception of philosophy as wholly and purely analytic.

As suggested, some of the philosophical issues raised in African thought are of great concern to people everywhere. Existentialism may not pass the rigorous test of analytic philosophers, but it surely has a better appeal to European thinkers and the whole populace than logicism or logical positivism. In modern philosophy all the philosophical thoughts rejected by analytic philosophers are still very much alive, and it seems to me that, even if there were life after death, philosophers would still continue to discuss the existence of God, being or non-being, why there is something rather than nothing, the mind-body problem, the meanings of life and death, human destiny, freedom and determinism, what is the essence of a person, and many more philosophical issues that seem to concern people everywhere. If there were to be any difference in future discussions of these philosophical issues, it certainly would not be in content or substance; perhaps it would be in a more philosophically sophisticated way of arguing than the contemporary discussions of these age-long philosophical issues. Perhaps an after-life is too long a time to conjecture. In another hundred years these philosophical issues would be discussed without loss of content, but most probably in a more sophisticated way.

Now, if the philosophical issues mentioned above are issues in both traditional and modern philosophical thought, then the current debate on African philosophy seems to me a misdirection of energy for, with the above kinds of philosophical issues in traditional thought taken as rough and ready materials, African philosophy could be done rather than endlessly debated. This is to say that since not all African traditional thought is unphilosophical, then some of it is philosophical. And if some African traditional thought is found perfectly comparable to what is admitted as part of Western philosophy, then Africans themselves must critically examine that thought, for in it there may exist other issues that are likely to enrich, rather than impoverish, our conception of philosophy. But the implication of the current debate on African philosophy is not that some African thought is philosophical, but that probably none is.

The less we recognize this implication as an important issue, the farther away are we from grasping the real point of the debate. And the closer we come to recognizing that not all African thoughts are unphilosophical and that not all philosophy is analytic philosophy the more we see the debate not only in pejorative terms, but as a self-defeatist mentality of a people whose official languages of philosophy are colonial. My own position is that it is not the case that all African thoughts are unphilosophical, just as we would say that not all Western thoughts are philosophical. And it is from the various philosophical issues in African thought which, in many respects, are found comparable to some of the philosophical issues in Western thought that a positive discussion on, and the teaching of, African philosophy may properly be started. From comparisons, different opinions would emerge and the difference, if any, between the underlying assumptions and issues in African philosophy and Western philosophy, may be discovered. Waiting for others to think for them is a typical problem among Africans. It is therefore my strong view that contemporary African philosophers should do the thinking themselves and not expect others to do it for them, as the colonial theologians, anthropologists, and ethnographers before them did for the whole of Africa.

Happily enough, not all contemporary African philosophers trained in the Western analytic tradition hold a negative view about African philosophy. There are many of them who think along the lines of Jean-Paul Lebeuf, a well-known ethnographer. In his speech to the First International Congress of Africanists held in Accra, Ghana, in 1962, Lebeuf made the following point in his survey of African philosophy:

> [Recent studies have shown] the existence [in African thought] of perfectly balanced metaphysical systems in which all the phenomena of the sensible world are bound together in harmony. . . . It cannot be said too often that the recording of these ontologies has rendered accessible a form of thinking which is as impeachable in its logic as cartesianism, although quite separate from it.[52]

Thus, the category of philosophers who see philosophical merit in African thought have not bothered themselves to join in the debate on whether or not there is African philosophy. Rather, they concentrate their efforts on research as well as writing on philosophical issues in African thought. Their efforts have yielded some positive results, as some of them have succeeded in applying their knowledge of Western philosophy to critical and even rigorous philosophical treatment of valuable African philosophical thought.

Some of the issues discussed are not only substantially philosophical, they are perennial issues in philosophy. It is to this category of contemporary African philosophers that the present writer belongs. My belief is that African philosophy can only be a subject of discussion if it has content in the form of philosophical thoughts, whether dogmatic or critical, speculative or empirical, true or false. In this way the contributions of

John Mbiti on African perception of the world of reality as well as his discussion on African concept of time provided useful and interesting materials for philosophical criticism. The same could be said of other African thoughts of which some contemporary African philosophers are giving philosophical analyses.[53] This surely is the way to begin, if we are to begin at all, to unravel the philosophical issues in some thoughts that otherwise might have been unphilosophically cast aside. If eventually an authentic African philosophy is to emerge, it will come from the criticisms and philosophical analyses of the writings of this category of contemporary African philosophers who, in the opinion of an American philosopher, may be said to represent the "spirit of progress" on the seemingly protracted debate on African philosophy.[54] This is why I call them the positive thinkers among the contemporary critical reorientationists in African philosophy. To them criticism and analysis are important tools but not a substitute for philosophy.

Perhaps I should have confessed that I was once a philosopher with a strictly analytic bent, and a great admirer of the now discredited school of Logical Positivism. But when it dawned on me that philosophy was not just logic and epistemology, or empirical science, I did some research on African thought and discovered that the Yoruba people of Nigeria had philosophical thoughts that were quite comparable to Western philosophical thoughts, particularly in the area of metaphysics. For instance, there is a remarkable similarity between Plato's doctrine and the Yoruba doctrine of the soul. I was, in fact, fortunate to have stumbled on a myth--the theory of seven heavens (orun meje)-- which is similar to Plato's myth of Er, by which the Yoruba, like Plato, justify the theory of the immortality and transmigration of the soul. In both there are indications that the soul is represented by the physical head as the seat of rational faculty and human knowledge, or what the Yoruba call wisdom (ogbon).[55]

In the modern identity theory of mind and body, the mind has been identified with brain states, particularly in Western philosophy and society. In this connection, it is pertinent to examine the theory from the point of view of an African metaphysical system. In his philosophical discussion on the human person and immortality in Ibo metaphysics, Richard Onwueme points out that "The theme of the human person and immortality, though a watershed of philosophical discussions in the past, has currently and forcefully become an issue in the face of modern materialism, and dehumanization."[56] But, by maintaining the irreducibility of the core of the human person (the soul) to a materialistic basis, he sees the Ibos as subscribing to a view of transcendence in their metaphysic of the human person, a transcendence which finds its expression in an egalitarian spirit of the society.

It can be argued that materialist or physicalist views are not adequate for a total conception of man; for man's aspirations, values, and relative achievements in the arts and science do not make full sense when discussed only from the backdrop of materialism and physicalism.[57]

Onwueme's position on Ibo metaphysics appears to me a good example of the rejection of the Western philosophers' identity or reductionist theory of the mind to the physical body by some African philosophers who have taken care to study and analyze the thought of the people of their own cultures.

The rejection of materialism is not a rejection of science but a rejection of science based purely on materialism at the expense of human values and personhood. A science based primarily on materialism looks at a person as if he had no soul or mind, that is, as an object other than a subject, to be conquered, controlled, or manipulated in much the same way science is used to conquer, control, or manipulate nature and all soulless automata. In fact, the dehumanizing effects of modern science and technology seem to strengthen the case for a philosophical rejection of materialism or physicalism which can be construed as an inhumane philosophical doctrine.

The Yoruba also have the concept of a person which, as in Western philosophy, consists of the mental and the physical elements known as soul (emi) and body (ara), with a clear indication as to how each came about.[58] One important difference between the Western and the Yoruba conceptions of a person is the complication the latter introduced to it. Although the Yoruba hold primarily to the dualistic conception of a person as mind and body, there is a third element of a person--ori (inner head)--as the bearer of human destiny.[59] Like the soul or mind, it is spiritual. In its relation to human personality it determines the destiny of man on earth. But, as Wande Abimbola puts it, a choice of good ori does not automatically lead to success. There is need for human effort to make a potentially good ori come to fruition.[60] Essential to human destiny, therefore, are character, effort, and appropriate sacrifice. Our analysis has shown that the introduction of human effort, character, and sacrifice or propitiation into the relation between ori and human destiny suggests an element of freedom, that is, freedom to make effort, make appropriate sacrifices and behave in certain ways in order to make one's potentially good choice of ori and destiny come to fruition. Because of this, we have seen that the Yoruba conception of destiny is not a strict one, like the kind described by Gilbert Ryle as "what is was to be."[61] The introduction of freedom to make efforts in order to make a destiny come into fruition thus allows for moral responsibility and punishment, in the same way freedom does to determinism. Fatalism, therefore, is not a viable concept in African philosophy. Since the Yoruba concept of ori is taken as an important element of human personality we see a tripartite conception of a person, with ori as the seat of human character and destiny, as it is in the concept of a person held by the Akan people of Ghana.[62] Both from oral evidence and from the work of indigenous African scholars there is evidence that Ifa remains one of the foundations of knowledge. In fact, our discussions above are from Ifa, which I have described in a paper read at the XVIIth World Congress of Philosophy in Montreal, Canada, in 1983, as a repository of knowledge.[63] Abimbola sees Ifa divination literature as a body of knowledge containing several branches. This shows that an important way of knowing in African philosophy, particularly among the Yoruba, is associated with the doctrine of Ifa which may be called Ifaology and compared with Laplace's omniscient intelligence.

Identified with the knowledge of all things, Ifa is described as an inexhaustible repository of knowledge (imo aimo tan). From this it is believed that what is known, and can be known, about the world is very small compared to what we can never know. Although we daily strive to acquire more knowledge through our study of Ifa, our knowledge of things around us and those beyond shall forever remain incomplete and imperfect. And because the knowledge gained from Ifa is a derivative one, that is not first hand knowledge as that possessed by Orunmila (the Yoruba creator god) himself, we have to content ourselves with probability which, for all practical purposes, serves the need although not the curiosity of human beings.

All this is by no means exhaustive of any discussion on the emerging field of African philosophy. While we refer to our examples as Yoruba thought or African philosophy, we do not mean it to be a group mind of the Yoruba people or the collective thought of all Africans. If this were to be so, we would have to say that Hume's skepticism or William James' pragmaticism is a "group mind" philosophy, a collective thought of the Scottish or American people. There are bound to be dissenting views or opposing philosophical systems, not only among different cultures of the world but even within one and the same culture.

Other important works on African philosophy consider the African foundations of Greek philosophy, time in Yoruba thought, cause and chance in Yoruba thought, and causal theory in Akwapim Akan philosophy.[64] My own writings, among others, discuss the Yoruba concept of a person and immortality of the soul. In recent years comparative study of African and Western philosophy has emerged.[65] In social and political philosophy, Chief Obafemi Awolowo of Nigeria has done a great deal of work, particularly in his philosophical discussion on "the regime of mental magnitude" as essential to African socialism.[66] Julius Nyerere has written on African socialism as "familyhood" or "community spirit," and Leopold Senghor on socialism and negritude. Nkrumah contributed his view on socialism as dialectic materialism, and his idea of freedom and unity.[67] These well-known African writers and statesmen are not just national ideologists as some contemporary African philosophers contend, but are African social and political philosophers whose original works are likely to confer on them the title of contemporary African philosophers.[68] If the works of these political philosophers were Western in origin, our Westernized African philosophers would not have labeled them as mere ideologies. Marxism is an ideology, a national ideology that hoped to drive its tentacles into other nations through indoctrination and brain-washing. Yet it is also a philosophy. And because it is a Western ideology it is accepted by our Westernized African philosophers and avidly read and studied as a political philosophy. Surely, this way of thinking is like George Orwell's system of doublethink, a system of thought that has no respectable place in philosophy. The effect of doublethink is mind control, and a mind that is controlled is one that has been manipulated or brain-washed into believing, if possible, that two plus two equals five, or, for the purpose of this paper, that the ideas of one's culture, however good, intellectually stimulating, and relevant to the survival of one's own culture, must be rejected or condemned in favor of the ideas of other cultures, however unsuitable and impracticable they may be to one's own culture.

From the foregoing I would like to make the following points. If the contemporary African philosophers had put their minds seriously to the question of African philosophy and probed further into some of the issues raised in African thought, they probably would have discovered some important materials from which African philosophical ideas could be drawn. At the very least such attempts could have set the stage for a more positive approach to the understanding of African philosophy and its possible incorporation into the philosophy syllabus used, not only by African departments of philosophy but by other philosophy departments in both East and West which may be interested in the development of African philosophy. In the United States of America, for instance, African philosophy is receiving some attention. A great deal of interest was shown in the subject during my one academic year at Ohio University, Athens, Ohio where, in the Spring Quarter of 1984, I initiated and taught African Philosophy as a full credit course in the Department of Philosophy. It is perhaps instructive to point out that the present writer's United States Fulbright Award was for further research in African Philosophy, as well as culture and traditional medicine. This award is a recognition of a positive approach to the emerging field of African philosophy and traditional medicine. As a part of my effort to upset the pattern of debate on the question of African philosophy and the controversial issue of traditional medicine the present work would serve a useful purpose. This effort can be seen as an attempt to improve people's understanding of African philosophy, and a further encouragement of the writing and teaching of African philosophy in both African and American universities.

From my own experience in the teaching of African philosophy to undergraduate and graduate students at Ohio

University and my public lectures on African philosophy at both
Ohio University and Central State University, Wilberforce, Ohio,
I have come to realize that more interest is generated in
positive discussion on African philosophy than a negative debate
on its existence, a debate that carries with it all the analytic
skills and conceptual rigor of an already familiar Western
philosophy. People want to know precisely what there is that
makes the difference between Western philosophical thought and
African philosophical thought, as it is already known between
Western and Asian or Oriental philosophical thought. Without
such knowledge there would be no distinction to be made between
the different philosophies and there would be no need for the
study of comparative philosophy. So, while a general debate on
the nature of African philosophy might prove a valuable exercise,
at least from the point of view of philosophical criticism, such
a debate is likely to cover up the more fundamental and fruitful
aspects of African philosophical thoughts that may be of interest
to Western and Oriental philosophers. Because some contemporary
African philosophers are perpetually looking at philosophy purely
from the Western analytic point of view I shall call them, for
want of a better name, "Euro-African Philosophers," with Euro-
African cultures and values as created by the circumstances of
language and their training in Western cultures and philosophy.
   I guess I also am one of them, if only by virtue of
language and training, and only trying to be different. But the
typical Euro-African philosophers are the contemporary African
philosophers who, because of their Western training, debate
endlessly as to whether or not there exists an African
philosophy, because they think that philosophy must be "rigorous"
or "exact" without showing precisely what they mean by
"rigorous," "exact," or "inexact" philosophy. But I believe that
the words rigorous and exact are more suited to mathematics than
to philosophy. Western philosophers know this. This view of
philosophy is drawing the hands of the clock backward as far as
the development of African philosophy is concerned. Even the
purely analytic conception of philosophy is no longer as popular
in Western philosophy as it used to be. That is why we recognize
metaphysical systems, existentialism, Confucianism, Buddhism;
that is why we recognize discussions on euthanasia, abortion,
feminism, sports, the endangered species, environmental pollution, being human in a technological age, the meaning of life,
the future of man, and applied philosophy, as relevant. They
probably are more engrossing and useful philosophical issues than
sterile discussions on the meaninglessness or absurdity of
concepts or propositions arrived at by means of logical or
linguistic analysis.[69] Incidentally, we do not require that
Chinese philosophy and Indian philosophy become analytic before
they are recognized as philosophies. African philosophy need not
be different. It is precisely from the recognition of each of
these as autonomous philosophies deserving academic respectability that the idea of comparative philosophy came into

being.[70] Philosophy is enriched, rather than impoverished, by such a comparative study of different philosophies and cultures. This must be so since no group of people possess, or ever claimed to possess, the exclusive preserve of knowledge and wisdom.

Who is an African Philosopher?

The questions are often asked: Is there an African philosopher? Who is an African philosopher? These questions often arise from the controversy as to whether or not there is African philosophy. The very controversy as to whether or not there is African philosophy is presupposed by the question as to whether or not there are African philosophers, at least in the sense in which we talk of Bertrand Russell as a British philosopher. This question springs from a wrong source, that is, the very controversy as to whether or not there is African philosophy; it is made to appear as if the answer to this question would decide the matter either way. But I would submit that the question should be raised from a different angle. From this angle the question becomes more meaningful, although more of a dilemma, for when raised we may not be able to describe an African as an African philosopher, irrespective of whether or not there is African philosophy. That is to say that even if African philosophy were now to be as fully developed as Western philosophy, it may very well be the case that there are no Africans to be known as African philosophers after all.

With the above preamble we shall now examine the question as to who really is an African philosopher, especially as the Africans trained in Western philosophy teach and write philosophy in Africa or abroad with the kind of technical competence acquired from their Western training, and not from training in African philosophy itself. In fact, those of us whom I call critical re-orientationists in philosophy are able to enter into the debate of whether or not there is African philosophy because we were already trained in the Western philosophical tradition. Here it must be noted that none of those (including the present writer) who have debated the question of African philosophy and taught courses in it had a formal training in African philosophy. This is to say that our critical re-orientation in African philosophy was brought about by our contact with, and training in, Western philosophy by which we are now able to criticize the first re-orientationists in philosophy, the colonial ethno-philosophers. Admittedly, contemporary African philosophers are talking, doing, writing, and teaching philosophy in a way quite different from their predecessors. But can they be called African philosophers, since their approach to philosophy, whether construed as Western or African, is essentially Western? The word "Western" here refers to England and France which have exerted their influence through their languages on our thinking and doing of both Western and African philosophy.

Practically all of the African philosophers had their training from the countries in which the language of their ex-colonial masters are spoken. Thus, philosophers from the former British colonies were trained especially in Britain, America, and Canada, and those from the former French colonies trained mainly in France. There are other colonial territories, but for our purpose the division into Anglophone and Francophone philosophical temperaments in Africa is sufficient. As would be expected, the majority of philosophers from Anglophone countries lean towards analytic philosophy, and the majority of philosophers from Francophone countries are greatly influenced by the philosophical temperaments in continental Europe. This situation is a healthful one for Africa as it may produce varied philosophical systems on the continent. But insofar as both Western and African philosophy is done, either in the English or French language in African countries, the question becomes even more pressing as to who really is an African philosopher.

Now, let us suppose you are an American philosopher. You and your African counterpart embark upon research in African philosophy. Because African philosophy is a relatively new field, neither of you had any formal training in the subject, and so it looks like a no-man's land. After some hard work, you write and teach African philosophy as competently as your African counterpart, a situation that might not have been easy for you if the language of African philosophy and instruction were to be African, rather than English. It must, of course, be noted that what we call language is inextricably connected with thought. The question can then be raised: Who, among the two scholars in African philosophy, is an African philosopher? What entitles one to be called an African philosopher? One might say that, first and foremost, it is being an African. But this could only be so if all that matters is nationality. In this case you, as an American, would not be an African philosopher. The best you could be in relation to your expertise in African philosophy would be "a philosopher of African philosophy." But this is no help since, whatever our African philosopher is, he is also a philosopher of African philosophy.

The problem of making a clear-cut distinction between an African who specializes in African philosophy and an American who specializes in the same subject rests on language. Because of the argument from language the problem may be pushed to the point in which the very idea of an indigenous African philosopher of any description may be completely eliminated from current philosophical discourse. This is to say that, in fact, an indigenous African philosopher may not exist, even if there is African philosophy! While African philosophers do write and teach African and Western philosophy in foreign languages, Western philosophers also write and teach African and Western philosophy but in their own languages which also are the languages of instruction in African universities. It is reasonable, there-fore, to argue that while they may claim to be African philosophers by virtue of the fact that their language is in fact the language of African philosophy, African philosophers who write and teach British or American philosophy cannot claim to be British or American philosophers because their adopted language is foreign to them.

We might extend this argument to cover an African who claims to be an African philosopher writing and teaching African philosophy. From this it might further be argued that whatever you call him, he is anything but an African philosopher. This is so because the langauge of African philosophy is not an African language. He certainly has his own language, an African language, which is not the language of African philosophy in Africa, Europe, or America. Who, therefore, is really an African philosopher? A Nigerian, a Ghanian, a Senegalese, a Briton, or an American? In spite of the obvious fact that he could not be called a British or an American philosopher, is there an African who may be called an African philosopher? I leave this puzzle for the reader to solve. I am sure some people would find my position on this issue very shocking, but I would maintain that, until African philosophy is written and taught in an African language, African philosophy may turn out in the future to be nothing but Western philosophy in African guise. As to who may be called an African philosopher, there may be none.

But perhaps this does not matter if only a distinction still could be made between African minds and thoughts and Western minds and thoughts, at least from the point of view of original languages and cultures. Since a people's culture and thoughts are better understood through the medium of their language as opposed to readings in translation, African philosophers could argue that they have an edge over those of their Western counterparts who do not understand native African languages. They might then argue that this circumstance naturally makes it easier for them to understand and interpret the cultures and philosophy of African people than it does Western philosophers. This is to say that Africans are in a better position to understand the cultures and thoughts of Africans and African philosophy.[71] After all, their minds are African although their thoughts have been greatly influenced by Western language and thought while that of the British or American philosophers are purely Western. Arguing that the Western philosopher understands only his native language while the African philosopher understands some Western languages in addition to his own African language, the African philosopher may then go on to reverse our earlier position and say that he is the one that could properly be called an African philosopher rather than the British or American who happens to do African philosophy without a proper understanding of African languages, thoughts, and cultures. By this the African could claim an advantage of first-hand knowledge of African cultures and philosophy over his Western counterpart to whom the "unspoken" and "unwritten" language of African philosophy is certainly a foreign language.[72]

56

scholar with a great anthropological bent, was the author of
Muntu "an outline of the New African Culture." For many years he
concerned himself with the study of African cultures, mainly
through literature and art. His book was written on a wide range
of new and traditional African thought as he found it in
religion, language, philosophy, art, music, and dance. He was
acquainted with Levy-Bruhl's work and was sometimes critical of
it. Jahn too approached African philosophy from the anthropological and ethnological points of view.

John Mbiti, a Kenyan, who could rightly be called a post-
colonial scholar, obtained his doctorate degree in theology from
the University of Cambridge, England. The author of the Concepts
of God in Africa, Mbiti's work on African philosophy is best
known in his well-known book African Religions and Philosophy.
For Mbiti the Africans live in a religious universe, meaning that
all the activities and thoughts of the Africans can be expressed
and understood, without remainder, from the point of view of
religions. This, of course, includes philosophical and
scientific activities. His overall picture of African religions
and philosophy seems to give the impression that there is a
diversity of religious beliefs in Africa, as well as different
philosophical systems. But since the belief in God is primary to
all African religions, and since the thoughts of Africans have
their mainspring in religions, it appears that Africans might
have a common philosophical system, for Mbiti thinks that
different philosophical systems among African people have not
been formulated. Mbiti's conception of philosophy as traditional
religions and culture puts him along side the colonial scholars
who approached the subject either from the religious, ethnological, or anthropological point of view. He makes easy references
to sociological and anthropological works, and his own view of
African philosophy is ethnophilosophy, a common philosophy
founded on the religious traditions of the people of Africa, all
of whom shared a common belief in a religious, nonscientific
universe.

Now, it appears that the interest of anthropologists and
theologians in African thought must be quite different from the
interest of the philosophers. An anthropologist may study
African thought as a social phenomenon while a theologian, like
Mbiti, may look at African thought from the point of view of
religion. In this way the anthropologists and theologians may
not have used the words "philosophy" and "philosophical,"
"logic," and "logical," in quite the same way philosophers use
the terms. But then it may be said that it is not the business
of anthropologists and theologians in their professional
capacities to criticize or pass judgment on African thought, on
particular concepts or passages of thought as philosophical or
nonphilosophical.[33] Philosophers throughout the ages believe
that philosophy can help us see the world and understand others'
unfamiliar ways of thinking as a way of enlarging our conceptions
of reality and the way other people think about it, from the

Chapter 4

THE SOCIAL AND POLITICAL PHILOSOPHY

OF OBAFEMI AWOLOWO

Philosophical inquiry concerning the principles of governing and social justice has been a concern from Plato to John Rawls. The history of Western philosophy is replete with theorists in social and political philosophy. In the West, the idea of government seems to rest on Hobbe's spirit of "covenant," Rousseau's "social contract" or John Locke's "government by consent." Two themes that run through all of these theories are democracy and justice. But since Plato, the concept of justice has defied accurate definition, while democracy, as "government of the people by the people and for the people," has remained so only on paper. Every society has its own system of values. This in turn so affects social organization that philosophers have produced a lot of work on social and political philosophy. But one observer noticed as follows: Humanity has never been able to discover a government that rules perfectly over all men and women even though philosophers and statesmen have penned millions of words about perfect government.[1]

The present writer has no record of ancient or medieval writers on African social and political philosophy. But we do know that prior to the advent of colonial rule, Africans did govern themselves and lived in harmony with nature.[2] Social and political organizations were sustained by traditional hierarchical systems of government. According to Abimbola, the Yoruba cosmos is based on a hierarchical order at the top of which sits <u>Olodumare</u> (Almighty God), assisted by the <u>orisa</u> (lesser gods). The next, in a descending order of importance, are the ancestors. All of the above are believed to be in heaven (<u>orun</u>). But the supreme power on earth belongs to an <u>Oba</u> (King) who represents the divine authority. As Abimbola writes:

> On earth, the power of <u>Oba</u> (king) is supreme over his subjects. The major Yoruba kings are believed to be directly descended from <u>Oduduwa</u>, the great mythical ancestor of the Yoruba. They therefore have divine authority. The <u>Oba</u> (kings) are assisted by a paraphernalia of town and village heads known as <u>baálè</u> who are in turn assisted by lineage or family

heads known as baálé. The baálé takes his decisions with approval of the household adults who are known as agba (elders). In this hierarchical order, children and young people occupy the lowest position. They have no authority whatsoever, and if they die before they become elders, they cannot become ancestors.[3]

The hierarchical structure of authority in Yoruba traditional society is presented in a rough sketch by Abimbola as follows:

Yoruba	English
Olodumare or Olorun	Almighty God
Òrìsà	Divinities
Òkú-òruń	Ancestors
Oba	King
Baálè	Village and town heads
Baálé	Household heads
Àgbà	Elders
Omodé	Children and young people

Traditional African society was as simple and humane as it was disciplined. Because of the communal life of African people, individualism, which plays a prominent role in the social and political structures of Western societies, is a remote concept. According to Julius Nyerere, individuals in a traditional African society are rich or poor according to whether the society is rich or poor.[4] This means that the poor cannot starve since they can depend on the wealth of the community of which they are members. Land, which is one source of wealth, is the gift of God and so belongs to all. For these reasons he sees capitalism as a remote concept in African traditional societies. Although one may want to call this traditional political value communism or communalism, Nyerere calls it socialism. The tendency, of course, is to avoid any association with the Marxist Communist doctrine. But the word "socialism" is also a foreign word, and may therefore have different meanings to different linguistic groups in Africa. Thus, while socialism could mean, in the African context, communal life such as living together in the spirit of love and brotherhood, it could also mean Ujamaa, "Familyhood" or "community spirit," as Nyerere puts it, or, in Leopold Senghor's sense, "community society" founded "on the general activity of the group."[5] Others, like Nkrumah, may construe it as "dialectical materialism" and Awolowo as "normative social science."[6]

As far as Nyerere is concerned, African socialism is rooted in the past. It looks at society as a family unit. Nyerere thinks that the foundation and the objective of African socialism and democracy is the extended family.[7] Senghor seems to have echoed Nyerere's opinion when he says that, because of its communal nature, Negro African society has traditionally been

60

socialistic, and this would seem to show the existence of socialism in Africa even before the coming of the Europeans.[8] The same spirit was shared by Nkrumah in his Consciencism. "African traditional society," says Nkrumah, "is communal, egalitarian, and humanistic." He then suggests, "if one seeks the socio-political ancestry of socialism, one must go to the African tradition."[9] Nyerere and Senghor not only believe in past traditional institutions but also advocate them as social and political goals. Nkrumah and Awolowo, however, had different views about the revival of past traditions for African socialist states. While Nkrumah thought that a return to the past would be tantamount to putting the hands of the clock back, he nevertheless saw the need to synthesize the spirit of traditional beliefs, social and political institutions, with the modern trends, but not as a complete revival of old traditional structures. As for Awolowo, his socialist belief is so different from the rest that his social and political philosophy deserves special attention. The rest of the discussion shall be devoted to Awolowo's social and political philosophy because of my acquaintance with him as a person and my familiarity with his work.

It appears that the idea of fashioning an African social and political philosophy based on traditional values is a good thing. If there is any objection against such an idea, it is for the obvious reason that, while an African social and political philosophy must take cognizance of an African traditional past, it cannot ignore the present and future. Beliefs, ideas, and people's experiences change with time, and if the human mind is an important instrument of social, political, and scientific change, Africans in the twentieth century cannot be expected to think only in terms of the past. It is in this respect that Nkrumah's position, which seems to suggest integration of traditional with modern values, could be seen as a viable alternative to Nyerere's and Senghor's pure traditionalism. But in spite of his claims, Senghor's views are rooted in the scientific methodology of Marxism. His rejection of doctrinaire Marxism is based on his belief that it is totally unrelated to the reality of African situations and society. From this point of view, Senghor's concept of African socialism became a matter of "integrating socialism with Negritude."[10] Whether an African socialism can be seen, or must be seen, in the light of Nkrumah's dialectic materialism is quite another question. An elaborate criticism of Nkrumah's position has been given by some contemporary African philosophers.[11] From the general confusion about the meaning of socialism in the African context, a general question arises as to what is meant by African socialism, and whether such a thing exists.

Like Nkrumah, Awolowo does not believe that an African socialism, or indeed any socialism, could be based on the traditional past. But unlike Nkrumah, Nyerere, and Senghor, Awolowo goes further to maintain that socialism cannot be said to have its roots in any region of the world. From this point of view all talk about fashioning a socialist doctrine that is distinctly African is out of the question. In his implicit criticism of Nyerere, Senghor, and Nkrumah, Awolowo argues, in The People's Republic, as follows:

> Socialism . . . as it is generally agreed by all socialists, is a normative science. Before any theory at all can answer to the name of science, it must be of universal application. If any principle is purely and strictly peculiar to a given institution, region, or state, it may be a custom, practice, or even a theory, but it certainly cannot lay claim to the status of science. Just as there can be no African ethics qua ethics as a science, or African logic, so there can be no African Socialism.[12]

In his Socialism in the Service of New Nigeria Awolowo declares that, as a normative science, socialism "sets the standards of human ends and social objectives which economic forces must serve, and prescribes the methods by which these forces may be controlled, directed, and channeled for the attainment of the declared ends and objectives."[13] But, as Awolowo observes, those Africans

> who have deep rooted prejudices against socialism . . . have adopted the ambivalent approach that whilst what they call the European type of socialism is a foreign philosophy, there is a kind of socialism which is native and indigenous to Africa. This is the so-called African Socialism which, according to them, is more suited to Africa than the so-called Russian or Chinese Socialism.[14]

Continuing his argument from the People's Republic Awolowo contends: "Those who have spoken of African Socialism, or Pragmatic African socialism, have fallen into three major errors."[15] First, the protagonists of African socialism, in Awolowo's view, have mistaken certain African customs and social practices like savings through Esusu (thrift-society), and the family or communal ownership of land, for socialism. His second argument, which is very important, runs as follows:

> In the second place, though there was, by and large, absence of greed in primitive African communities for material acquisitions and extensive ownership of private properties, this, in our view, was not due to any adherence to the principles of socialism of which they were never conscious, but rather to insuperable physical obstacles to such acquisitions and ownership. In the absence of adequate and efficient

62

communications; in the midst of incessant inter-tribal and internecine wars, with their attendant grave insecurity to property and life; and in the absence of portable and durable means of exchange which, apart from anything else, could serve as store of value, the desire and the greed to accumulate the things of this world were reduced to the barest minimum.[16]

In support of the above Awolowo gave a brilliant example from his own culture. For instance, before the advent of the British, the traditional medium of exchange was the cowry shell. Twenty thousand shells of cowries make "one sack," an equivalent of five shillings, which is less than sixty cents. At that time it took an able-bodied person to carry one sack, while a man was considered wealthy if he had just one sack in his possession. Anyone who possessed fifty sacks, that is, twelve and a half pound sterling or about twenty dollars, was a millionaire. It would require fifty able-bodied persons to carry this monetary wealth in the case of an emergency.

In the circumstances, therefore, it would be madness for anyone to possess too much of either perishable farm products, or unwieldy cowries. [However], the invention of an easily portable and durable medium of exchange which, by itself, also has value, marked the beginning of excessive greed in the acquisition of material wealth. The improvement of communications also helped to fan this anti-social passion beyond all imaginable proportions.
As long as the barriers to extensive ownership of private property existed, the Africans, like all primitive peoples in other parts of the world, lived a life of simplicity and contentment, which was comparatively free from the greed and naked self-interest that are prevalent in a capitalist society. But as soon as these barriers were removed and a money economy was introduced, coupled with improve-ment in communications, the passion for the greedy accumulations of wealth became as sharp, venomous and devastating in the Africans as in the other human inhabitants of the globe.[17]

Only a little reflection would show the relevance of Awolowo's argument to present-day African society. All the changes he has talked about have made it possible for people to amass wealth, even to unreasonable proportions, in many African countries. If it took fifty men to carry twenty dollars before the advent of the British in Nigeria, how many men would Nigerian politicians, some of whom amassed up to one billion dollars from the public treasury, need in order to carry their monetary wealth from one place to another? About fifty million people, more than half the entire population of Nigeria, more than four times the population of Ghana, and twice the population of Canada! Modern technology, including printing and communication, has certainly increased the human passion for greed, as I have noted elsewhere.[18] While it would require fifty million people to effect the transfer of one billion dollars in cowries from one place to the other in the past, the entire sum of money can now be transferred on a piece of paper from one part of the country to another.

In his objection to Nkrumah's concept of pragmatic African socialism Awolowo argues that in pragmatic socialism there is an obvious confusion, especially between ends, on the one hand, and the method of socialist approach, on the other.

> Viewed from any standpoint, whether it is the Marxist, the Maoist, the Titoist, or our own brand of socialism, the normative social objectives are the same, whilst the methods of approach are conspicuously different from one another. If circumstances so dictate, a pragmatic approach to the attainment of socialist goals in a particular country may be adopted. But the fact that a particular tactic is adopted does not in any way alter the fundamental ends, or thereby necessarily make such ends pragmatic in themselves.[19]

In the light of the above arguments Awolowo concludes his criticism of the concept of African socialism or pragmatic African socialism by pointing out that, in the above sense, democratic socialism is on the same footing as pragmatic socialism but with an important difference:

> While pragmatic socialism may be democratic or otherwise in its approach to socialism, democratic socialism must of necessity be democratic in its methods--at any rate, in the manner of its deep contemplation and actual planning by its adherents.[20]

As we shall see later, one big difference between Awolowo's socialism and a Marxist socialism is in methodology. In his critique of pragmatic African socialism, he makes it clear that the end does not always justify the means. If socialism can be attained by violent or democratic means, Awolowo rejects the former as a means to a socialist end, whether in Africa, Europe, America, Russia, or Asia. For this reason he propounded a philosophical doctrine on which his socialism rests. It is a doctrine which is quite consistent with his universalist view of socialism.

The Regime of Mental Magnitude: The Philosophical
Foundation of Awolowo's Socialism

In chapter one of this book we talked about changes of beliefs and customs from the Middle Ages through belief in the divine right of kings and the rejection of this belief, and from the empiricist philosophy of Francis Bacon, John Locke, and Thomas Hobbes to modern technological cultures. We also provided an African (the Yoruba) traditional structure of society and its government, which had also a belief in the divine authority of <u>obas</u> (kings), as given by Abimbola. Although the rejection of the divine right of kings in Western societies was based on empiricism which led to skepticism about God from whom kings were said to have derived their earthly power and supremacy over their subjects, rejection of divine right in Africa was somewhat different. The kind of social system discussed by Abimbola as given earlier no longer obtains in Nigeria, and I am sure in no other parts of Africa. One may believe in God and yet reject the idea of the divine right of kings. Such a rejection would be based on a different belief--the inadequacy of man and the weaknesses of human nature.

In traditional African society, the <u>oba</u> (king) was an extremely powerful person. He had the power of life and death, and could do whatever pleased him in the society. Above all, there was not the required training of the mind (through education) to prepare an <u>oba</u> for the art of governance, although people lived peacefully together in a community, perhaps for the reasons or circumstances already given above by Awolowo. The conditions for greed and excessive material acquisition were not there. For this reason it would appear awkward in modern time to advocate a return to an African traditional system of social organization to be known as African socialism. However, whatever sense of socialism is meant, there is one thing that is generally accepted by African social and political philosophers. Whether one talks of socialism as familyhood, community spirit, dialectical materialism, or normative science, one thing that seems to be essential to all is living together in a society in a spirit of love and brotherhood. That should be the universal objective of socialism in any region of the world. But for this objective to be met, certain antecedent conditions must be met, judging from the nature of human beings everywhere and at any time.

In his paper on Awolowo's political philosophy, Nwanwene highlights five imperialisms identified by Awolowo which are marked down for elimination on the African continent.

1. Imperialism of the rule imposed by one State on another
2. Imperialism of ignorance
3. Imperialism of disease and want
4. Imperialism of capitalism

5. Imperialism of what Kant calls "The tyranny of the flesh.[21]

According to Nwanwene, Awolowo's political philosophy is aimed at the building of a state free of all these imperialisms. Awolowo's Path to Nigerian Freedom (1966) took care of the liquidation of the first imperialism, the rule of one nation by another which "cannot be justified by any standard of morality" and which he sees as a "symbolical oppression, tyranny, and the ruthless exploitation of the masses by the privileged and powerful few."[22] Although Awolowo's conception of liberty, defined as "a state of freedom," is not an absolute one, he believes that political freedom is a necessary although not a sufficient condition for the termination of economic enslavement to colonial powers.[23] To be politically free, an African state must also be economically free. Since Awolowo does not believe in absolute freedom, no African state can achieve absolute economic freedom. Economic freedom here must therefore be measured in terms of economic strength, not just of some individuals, but of the state as a whole. Therefore, some of Awolowo's grievances against capitalism are greed and naked selfishness seen as the predominant motivation of capitalism "which is bound to generate secular social disequilibrium in the society in which it is operative." It appears from this point of view that, although Awolowo has some good things to say about capitalism, economic exploitation would be a vice in a capitalist society.[24] It is not the purpose of this paper to discuss Awolowo's criticism of capitalism which, in Nkrumah's assessment, is "the gentlemen's method of stealing," and which Nyerere describes as a wrong attitude of mind which colonialism brought to Africa.[25] What are being rejected in capitalism, therefore, are its methods and values.

In rejecting the capitalist values of greed, selfishness, and exploitation, Awolowo makes it clear that his own socialism is neither communism nor a form of Marxism. "Our concept of socialism is entirely different from communism and the Marxian concept of socialism."[26] Sekou Toure had believed that without being communists the analytical qualities of Marxism and the organization of the people were methods well suited for Africa.[27] Awolowo would not agree with this view, especially from the point of view of the methods of achieving socialism in Africa or elsewhere. Therefore, Awolowo rejected four Communist or Marxist ideas.

(i) Violence. The elimination by force or violence, if necessary, of the capitalist class by the Proletariat (mainly workers) who will assume all responsibilities of administration of government under the Dictatorship of the Proletariat. This is to say that the end justifies the means, and it does not matter what means are used, including violence; any means must be used towards the

66

desired end. For Awolowo, force or violence is not inevitable in the evolution of a socialist state or in its governance.[28]

(ii) One Party State. In his address to the students' Parliament at Ahmadu Bello University, Zaria, Nigeria, on "Representative Government: Theory and Practice," on Friday, 6th December 1975, Awolowo spoke on the topic "Democracy is the best form of Government."[29] Because of his love for democracy he regarded as evil the one party system of government which he saw as a sine qua non to the Marxist-Leninist socialism. A one-party system of government, whether military or civilian, is evil because it is unelected. Thus Awolowo's political philosophy is quite in agreement with John Locke's "government by consent" as expounded in his Second Treatise on Government. Awolowo's socialism is therefore Democratic Socialism.[30]

(iii) Lack of Personal Freedom. Awolowo rejected the fundamental tenet of Marxism that there can be no personal freedom in a socialist state. Communist or Marxist socialism aims at fashioning a classless society where the need of every individual would be catered for. This is a noble aim, but, in practice, the means of its achievement may involve the curtailment of individual freedom and liberty, thus leading to totalitarianism. Awolowo believes that from the rational point of view, socialism, as a means of achieving an egalitarian society, does not involve the curtailment of personal freedom. Thus one of the slogans of his political party, the then Action Group, was "Life more Abundant, Freedom for all," including freedom of the Press.[31]

(iv) Intolerance of Religion. An important deviation from Marxist socialism in Awolowo's concept of socialism is his view on religion, especially his fervent belief in God and Jesus Christ, and in Mohammed as the chosen prophet of Allah.[32] Hegel, Marx, and Lenin were complete atheists. For Hegel, the State and not the unseen God was the proper object of worship. Hegel's belief is regarded as the fountainhead of Fascism and Nazism. Lenin, the author of the dictatorship of the Party, once emphasized that communism would never succeed until "the myth of God is removed from the minds of men," while Marx regarded religion as "the opium of the masses" for which reason eternal truths, all religion and morality are to be abolished.[33] Awolowo, in contrast, believes that socialism and all the great religions, including Christianity and Islam, have the same objectives.[34] Earlier in the defence of his socialist approach Awolowo had given what must be regarded as the statement of these objectives as follows:

We declare that the aims of socialism are social justice and equality, and a state of affairs in which the resources provided by Nature belong to all the

citizens equally, and the products of the union of
land and labour are appropriated to labour of all
gradations and skills through the media of good
wages, respectable standards of living, abolition of
unemployment, free provision of social amenities such
as education, health, etc.[35]

With the exception of Nkrumah's socialism as non-theistic
dialectical materialism, there is no evidence that other African
social and political philosophers who wrote on socialism were
intolerant of religion.[36] As Nyerere puts it:

Socialism [African] does not demand from its
followers that they become atheists . . . there is
no contradiction between socialism and Christianity,
Islam or other religions which accept the equality of
men.[37]

From the above it is clear that the political philosophy of
Awolowo, commonly known as Awoism, is in conflict with the
orthodox or classical views of Marxist-Leninist communism or
socialism.

By far the most important issue in Awolowo's concept of
socialism is his philosophical doctrine of mental magnitude.[38]
On it rests the essential differences between Marxist socialism
as stated above and his rejection of the concept of African
socialism or Pragmatic African socialism as stated earlier.
Along with Nyerere, Awolowo believes that socialism like
democracy is an attitude of the mind, but not this or that mind,
relative to this or that region of the world; it is an attitude
imposed by rational mind in general.[39] In his "Ideological Re-
appraisal," the fourth lecture in The Problems of Africa, Awolowo
sees human beings as the instrument of social, political,
economic, and scientific changes. "Man is the sole creative and
purposive dynamic in nature: everything else by comparison is in
a state of inertia."[40] He sees man as more than just his
physical elements, "an animated lump of earth mixed with water."
He is more than this.

He is dual in nature: part animal, part God; part
conscious, part subconscious; part body, part mind.
He is infinitely superior to all other living beings
in the animal kingdom. According to Christian
ontology, God breathed into his nostrils and man
became a living soul. The living soul is housed in
the shell called body and this soul can only be the
same in kind and quality, though not in degree, as
the soul or Infinite Intelligence which pervades the
universe and animates man. Looked at in this
edifying way, every man, without exception, is a
potential genius.[41]

68

From this, Awolowo goes on, like Plato, to relate the organs of the body to those of a state. The quality of a state is that of its citizens.

However, it is in the People's Republic that Awolowo spells out his doctrine of mental magnitude. The whole of chapter nine of this book is devoted to it, and its sole object, as it is in the other chapters, are human beings. The basis of Awolowo's discussion is his philosophy of man, with some ideas borrowed from psychology. He argues that

> Proper knowledge of man, and a thorough appreciation and competent application of the principles which must govern his physical culture, his mental development, and his spiritual self-realization, is indispensable to any efforts for promoting and guaranteeing his general well-being and happiness.[42]

Like Descartes and other dualists and especially the Yoruba in their dualist conception of a person Awolowo believes in body and soul. As a physical being, man has a body with various organs, nerves, tissues, and innumerable cells. He is endowed with the faculties of sight, hearing, touch, taste, and smell. Awolowo seems to believe in innate ideas when he suggests that human beings also have innate dispositions, called instincts, endowed by Nature at birth. But there are significant differences between the senses on the one hand, and instincts on the other.

> The physical senses are man's instruments for observation, analysis, judgment, reflection, and reason. The instincts, on the other hand, predispose him to emotionalism and impulsiveness. The employment of any of the senses is a cognitive act: that is, a deliberate act of the will. Whereas feelings which arise from instinctive dispositions are independent of man's will. The feeling of hunger, for instance, is independent of man's will. Whether he likes it or not, when the previous meal is fully digested and his stomach is empty, he will experience the feeling of hunger. If he is affronted or insulted, his instinct of combat, coupled with the emotion of anger, is aroused independently of his will. On reflection, that is, on the application of his faculty of reason, he may bring his will-power to bear to subdue his emotion and curb his predisposition to combat. In this instance, his faculty of sight probably tells him that the person who causes him offence is too big for him to tackle successfully, or too small to contend with reasonably. If he suddenly found his child trapped in a burning house, his immediate reaction, arising from the parental instinct, would be to run to his rescue

69

without thought of his own safety. On reflection, his reason might direct him not to make the plunge, lest he and the child perish, and his other children should become orphans into the bargain.[43]

From the above explanation Awolowo contends that the five physical senses are rational and objective while the instincts are the seats of all man's emotions. Emotions can be regarded as either positive and good, such as emotions of curiosity, sexual feeling toward one's wife or intended wife, escape in the face of actual danger, construction, creativeness or productivity, laughter, and acquisition for the satisfaction of one's needs. Others can be negative and dangerous, such as anger, rage, fury, pugnacity, aggression, resentment, hate, fear, envy, jealousy, selfishness, and acquisition for the purpose of hoarding and self-display. From the epistemological point of view Awolowo seems to have taken both physical senses and the human innate instincts as functions of the human mind. The senses are associated with the faculties of reason by reflection, and capable of controlling the excesses of emotions, if only human beings have the will to do so. But the senses by themselves are not enough, as the sense faculties are more often than not defective, while the instincts almost always tend to overplay their parts. From this, Awolowo infers the following:

If all the organs in man, together with the five senses and all the instincts, are balanced and functioning normally and harmoniously as Nature intends them to, there will be no such things as negative emotions like those we have just mentioned. Murder, and all forms of crime, will be non-existent. Man would then live a full, happy, glorious, and triumphant life.[44]

The harmonious functioning of the five senses and instincts is the duty of man's critical faculty, an aspect of the human mind that aids the senses in reflection and rational judgment, acting, so to say, as the central coordinating authority, discriminating and separating defective perceptions of reality from good ones, as well as positive from negative emotions. Here comes the central point in Awolowo's doctrine. Having located the seat of the sense faculties in the brain, he sees the willingness or refusal to employ them correctly as an act of will, a function of the human mind. At this stage Awolowo entered into the Platonic division of the human mind into the rational and appetitive, the former governed by reason, the latter by instinct or mere desire. Since one of the human instincts is social, and since this is the essential aspect of society and its governance, any society must be analyzed from the point of view of human beings as its irreducible unit. In this respect Awolowo's introduction of mental magnitude puts people at the center of social, political, economic, and scientific changes. It leads to a profound metaphysical analysis of human beings in relation to mental and physical capabilities in society.

More than this, it shows the superiority of mind over the body. Like Socrates and Plato in the Republic, Awolowo believes that the body or appetite must be governed by the mind, and takes it to be "an immutable law" that the mind commands while the body obeys. The mind is the seat of reason while the body is the seat of mere appetite, instinct, desire, and the negative emotions. It is not at all clear here whether by "reason" Awolowo means the sense faculties which to him are faculties of reasoning, or pure reason, such as Kant takes to be completely independent of the senses. The latter will be what Plato means by the rational part of the soul. The relation of reason to the human mind is not clear in Awolowo's analysis. However, from the way he uses the word "reason" the senses act as the sources of materials on which the human mind acts upon by introspection and reflection, comparing and contrasting as well as discriminating one from the other in order to make rational judgment. Reason, in this sense, is a critical faculty of the mind and must be placed above physical desires. As a thought process, it is something quite above ordinary perception. As Awolowo himself argues, it is thinking that distinguishes human beings from the other animals and makes them the image of God which he usually refers to as the universal mind. Perception is not to be equated with thinking:

> Mere perception is not thinking; the lower animals do this too. But if perception is purposive, then it ranks as thinking. Because in that case, it becomes only the first stage to apperceptual, conceptual, and ideational thinking, each of which is necessarily purposive. Desultory use of the mind, wild imaginings and day-dreamings are not thinking. Even children are more adept in all these than adults. At best these are sheer conceptual or ideational dissipations.[45]

From the above description of the thought process Awolowo believes that only a very few people do real thinking, while the majority do very little thinking, if any at all. He, however, gives a more rigorous account of reasoning as a process of thought in inductive-hypothetico-deductive terms in the following argument:

> It involves observation, collection of data and materials, analysis, synthesis, and reasoned deduction. [It] demands attention, contemplation, idealization, visualization, and reflection. It needs rigorous training, self-discipline, and self-knowledge. It has tremendous rewards: self-conquest,

self-improvement, self-realization, and victory over
environment and heredity are some of them.[46]

Man, he argues, has used his reasoning faculty to conquer his
environment. He has used it to eliminate or minimize the adverse
effects of the weather by means of suitably and scientifically
regulated food, clothing, and housing.

As for the physical surroundings, he can make them do
his bidding. If they are inadequate for his purpose,
he must supply their deficiencies; if they are
hostile, he must subdue them. Man is not born to
grope in the face of adverse environmental circumstances and conditions: he is ordained, and endowed
with the capacity, to comprehend the universe,
conquer his immediate surroundings, and rule the
world. But first, he must understand the world and
all its phenomena: he must do so systematically and
scientifically.[47]

Awolowo looks at society from this scientific point of
view. It appears to him, as it did to John Stuart Mill, that the
same knowledge which has been used to conquer nature to man's
advantage can be used in fashioning a good, just, and egalitarian
society.[48] He agrees with Aristotle's definition of man as
political animal or a gregarious and social animal. Because of
this, man must learn to live "amicably and harmoniously" with his
family and other members of his society. "If the society or any
of its institutions is primitive, backward, underdeveloped,
oppressive, or evil in any sense, he has a duty to himself and to
the society at large to improve it."[49] Since every person is a
potential genius, each one of us can diligently improve society
and the world. But first, we must all understand our physical
and social surroundings, and in order to understand, we must be
properly educated:

For it is only when he knows the law that governs the
universe of which the world in which he lives is
part, the rigours of Nature, and the abberations of
society, and can, with the necessary physical,
mental, and spiritual equipment, devise ways and
means to temper and humanize them to satisfy his
sublimated yearnings, that man can hope to live a
full and happy life.[50]

Ideally for Awolowo, as it is for Plato, reason then should
govern the appetitive aspect of human beings, for it is the
source of true knowledge. But when reason is dethroned for
appetite, mediocrity, desire, and negative emotions, the
inevitable result is the corruption of the mind which in turn
leads to all sorts of evils, such as greed, bribery, nepotism,

abuse and misuse of office and power, stealing, cheating, collusion with contractors for the purpose of stealing public funds, smuggling, violation of the laws of the land, violence, and gangsterism--all of which flourish in Nigeria, Awolowo's land of birth. All these, he explains, flow from the negative emotions, the appetitive aspect of man, his desire for physical pleasure. He would claim that his analysis is not restricted to any particular society, but to every human society. Evils are to be combatted in African socialist states as much as in any socialist state. Again, while these evils are associated with capitalism which, from the point of view of greed, acquisitiveness and exploitation is not restricted to any region of the world, they are not, and cannot, be associated with socialism under the regime of mental magnitude. It is from this point of view that he sees socialism as a normative theory having its roots in the human mind, as derived from the universal mind or God, and not from an African traditional social system.

Socialism has one purpose: the evolvement of an egalitarian society. Therefore, for the purpose of effective governance and a just and egalitarian society those who aspire to leadership must be those who are ruled by reason rather than appetite. They must be mentally equipped, for a leader without a sound and disciplined mind cannot pilot the ship of state to success. If he succeeds at all, he can only do so at the expense of social and economic disequilibrium. For a leader to make a success of his difficult assignment of governing a state, he should possess comprehension, mental magnitude, and spiritual depth. Because mental magnitude and spiritual depth are used in a synonymous sense by Awolowo, the latter will be discussed shortly. By comprehension Awolowo means "the ability of a man to appreciate and grasp the salient details, as well as most of the practical and temporal implications, of a given problem or situation."[51] A leader possessed with mental magnitude is a leader always in control of himself. In order to be masters of others and able to instill the necessary discipline in their minds, one has to be self-disciplined, master of oneself.

> Men of affairs and wisdom everywhere are unanimous in the view that only "those who are masters of themselves become masters of others." Indeed Aristotle has said it, with the authority of one of the greatest and wisest men that ever lived: "Let him that would move the world first move himself." And if we may, in passing, adapt Aristotle's words, we should say: "Let the country that would lead Africa, first lead itself out of its own domestic mess."[52]

While Awolowo's argument may be seen as specifically directed against an African system of government and its leaders, particularly Nigeria, he sees his view as universal, and applicable to any leader or system of government whose aim is to

73

achieve a socialist state. In this connection, he considers it a settled opinion that the tyranny of the flesh is the worst of all the tyrannies. For this reason, only those who have subdued this tyranny can successfully lead others out of the bondage of ignorance, poverty, and disease. In short, "good leadership involves self-conquest; and self-conquest is attainable only by cultivating, as a first major step, what some applied psychologists have termed 'the regime of mental magnitude.'" In explaining how the regime of mental magnitude is cultivated, Awolowo reveals his kind of stoic philosophy which, for a leader, requires self-mortification of the highest order.

> In plain language, the regime of mental magnitude is cultivated when we are sexually continent, abstentious in food, abstain totally from alcoholic beverage and tobacco, and completely vanquish the emotions of greed and fear.[53]

He sees the above as the underlying causes of the maladies which afflicted Nigeria under the First Republic.

How is a regime of mental magnitude to be cultivated? From Awolowo's point of view it is through training of the mind. An essential tool for that purpose is <u>education</u>. It is for this reason that Awolowo always makes as the first cardinal program of his political manifesto for a socialist state free education at all levels, from kindergarten to the university, as a way of combating ignorance. He believes that a universal education of the citizens will lay a solid foundation not only for future social and economic progress, but also for political stability. It would also help in eliminating what he calls the imperialism of the local caesars, for "a truly educated citizenry is, in my view, one of the most powerful deterrents to dictatorship, oligarchy, and feudal autocracy."[54]

Now, what does Awolowo mean by education? He has this to say:

> The cardinal aim of education is not, as is popularly but narrowly conceived, to teach a man to read and write, to acquire a profession, to master a vocation, or to be versed in the liberal arts. All these are only means to the end of true education, which is to help a man to live a full, happy, and triumphant life. In other words, any system of education which does not help a man to have a healthy and sound body, an alert brain, and balanced and disciplined instinctive urges, is both misconceived and dangerous.[55]

From this Awolowo maintains, quite emphatically, that the aim of education should now be clear to all. "It is to make it possible for man's physical organs as well as his instincts to function normally, positively, and harmoniously."[56]

Since Awolowo is a dualist with respect to the mind and body, he believes in the relationship between both, although, like Descartes, he fails to explain how they are related. While Descartes locates the connection of mind and body in the pineal gland, Awolowo seems to have located it in the whole of the human brain. But, as human beings possess both mind and body, there is, in Awolowo's view, the need to care for both. Thus while the training of the mind is to be taken care of by free education at all levels, there must also be free medical care for all, as a way of combating diseases. In medical care Awolowo includes the preventive aspect which he sees in such things as good and clean environment, clear and clean water, good and nourishing food, and physical exercise. And by "exercise" he means physical, mental, and spiritual exercise. "The only way to exercise the brain is to study," and to do so "persistently and reasonably rigorously." A better education can also help people to live a good and healthy life. Thus, like Plato, Awolowo believes in the saying: "A sound mind in a sound body." It is when these two are achieved that human beings, as instruments of social, economic, political, and scientific changes, can contribute to a nation's development. What is more, it is only after the above two cardinal programs are settled that the third and fourth of Awolowo's programs for a socialist state naturally fall into place, that is, gainful employment and integrated rural development.[57]

What Awolowo's theory of mental magnitude shows is the importance of the state of mind on physical well-being, and in the development of a socialist state that is free from greed, selfishness, social injustice, and all sorts of dehumanizing activities that go along with desires for political and economic powers, oppression, curtailment of human freedom, and erosion of privacy. As he often says, a good leader would have nothing to hide, and nothing to fear, including criticism. A good purposive government is the most difficult to overthrow by force. He holds the view that, from the point of view of the regime of mental magnitude, the Presidency of Nigeria is not an office of pleasure. When all this is translated into his theory of a socialist state it shows the importance of the person of reason in the management of the affairs of a nation. And it is only when this is applicable to any socialist state in Africa that it can boast of a regime of mental magnitude. Some interesting implications of Awolowo's regime of mental magnitude is that socialism cannot be said to have purely African roots. It has its roots in the very nature of human beings as political animals. And that if socialism is to have any meaning in Africa, and if Africans are to govern themselves well in socialist states, they must cultivate the regime of mental magnitude lest their self rule after independence turns out to be a curse.

The Universal Mind

No discussion on Awolowo's socialism is complete without the religious foundation of his principle of human association. As we said earlier, nearly all of the African social and political philosophers believe in God which they found not to be inconsistent with the socialist ideals. But Awolowo's socialism goes further than just believing in God. God has an important role to play in his brand of socialism. Before his preface to The People's Republic Awolowo gives two important quotations from the New English Bible (Matthew 19: 16-24 and 22: 36-40). The most important quotation which Awolowo uses as a religious foundation of his socialist doctrine is the second of the above quotations.

> Love the Lord your God with all your heart, with all your soul, with all your mind. That is the greatest commandment. It comes first. The second is like it: "Love your neighbour as thyself." Everything in the Law and the prophets hangs on these two commandments.

We have been told that in order to rule effectively a leader must cultivate the regime of mental magnitude with all its stoic discipline. But this is not enough. He must also possess spiritual depth. That is, in addition to mental magnitude,

> the classic injunction known as 'the golden rule,' handed down by the greatest teacher and prophet of all times, should be most scrupulously observed. It reads: "Always treat others as you would like them to treat you: that is the Law and the prophets. . . ."[58]

Thus a new dimension introduced to his socialism is the religious concept of love.

> The touchstone of what is good, be it thought, or words, or action, is LOVE. We are to love our neighbours as ourselves. "That is the law and the prophets." Anything therefore - any thought or word or action - which falls short of LOVE is evil, and holds within itself the germ of its own eventual and inevitable destruction.[59]

As human beings are made in the image of God, so is their mind. But to God belongs the universal mind. "For ourselves, we believe in God, and believe that he is the creator of the universe. He is the UNIVERSAL MIND which permeates and pervades all things" (emphasis mine). Elsewhere in his book Awolowo maintains that the universal mind, which is latent everywhere, "is both immanent and transcendent. It is omnipotent, omniscient, and omnipresent."[60] It is from this universal mind that

76

the idea of love permeates the mind of man, and it is when man possesses this religious love, and acts in accordance with the Christian injunction "love thy neighbour as thyself," that he truly possesses spiritual depth and mental magnitude. It is the only insurance against greed or naked-selfishness which is the essence and predominant motivation of capitalism. The inference now is that only leaders possessed with mental magnitude and spiritual depth can bring about the existence of a true socialist state. In such a state, every citizen will also possess the same mental magnitude and spiritual depth as possessed by its leaders. Professor Oluwasanmi is right when he writes: "Socialism and mental magnitude are the two main pillars upon which rests Awolowo's universe of ideas and actions. Both are only two sides of the same coin."[61]

The relation of the universal mind to mental magnitude is obvious from Awolowo's perception of man as the image of God. The regime of mental magnitude can therefore be regarded as the philosophical foundation of Awolowo's doctrine, the alpha and omega of his social and political philosophy of Awoism. With the present political climate in Africa, and Nigeria in particular, Awolowo has defended his position with conviction, a certain expenditure of mental energy and personal authority. Our task in this book has been rendered less difficult by our reliance to a large extent on the mental energy and personal authority of Awolowo as demonstrated in our copious references to his teachings, speeches, and several of his well-known works on social and political philosophy.

Because he was never a Nigerian head of state, Awolowo did not receive the same attention given to Nyerere, Nkrumah, Senghor, and others who, at one time or the other, had the opportunity to put into practice their respective doctrines. It was only as Premier of Western Nigeria (1954-1959) that Awolowo had the opportunity to put into practice his own idea of a socialist state with the successful introduction for the first time in Nigeria of free primary education, free medical care for all, employment opportunity and a minimum wage for workers, the religious program now known as the Pilgrims Welfare Board and many others.[62] In 1979, his socialist program was broadened to include free education at all levels, full employment, and integrated rural development known as optimum communities (OPTICOMS). Because of his abstract philosophical doctrine of mental magnitude which, as in Plato's Republic, subjects human appetite or desire to reason in social, political, and economic behavior, and especially his denunciation of what he calls the "tyranny of the flesh," his political ideas were not popular among wealthy Nigerians. His ideas, however, had the support of the rank and file of intellectuals, students, workers, and the poor. The reason for the support of Awolowo's doctrine by the above groups of people, I believe, is also well-known. As one commentator puts it:

in a polity crying for an upright, purposive and responsive leadership, in a state suffering from and wallowing in poverty in the midst of plenty, and, in fact, in any underdeveloped state, the great doctrine of Awoism becomes a necessary philosophical armour for the leaders. No doubt Nigeria is today a sick baby, requiring urgent surgical attention. The diagnosis reveals <u>Indiscipline</u> and <u>Complete Misdirection</u>, and the panacea is proper leadership armed with the dual weapons of knowledge and discipline.[63]

A question that is likely to emerge from the above is why, in spite of the popularity of his numerous writings and socialist programs, Awolowo has failed to rule Nigeria, the most populous and richest of the countries in Black Africa? Political observers have attributed this failure to election malpractices by his wealthy political opponents which, from the philosophical point of view, can be attributed to his tough demand for stoic discipline, and his strict and rigid demand for a regime of mental magnitude. As one writer sees it,

his (Awolowo's) insistence on inflexible principles seems to have been the main reason why he failed either to capture or share federal power during the pre-1966 Nigeria. As a maximizer he could not make sufficient concessions which would earn him a piece of national political power. His rigid ideological orientations made collaboration with other more flexible political parties difficult and earned him their hostilities.[64]

While the above seems to me a correct assessment of Awolowo's operational codes as greatly influenced by his doctrine of mental magnitude, as a Platonic idealist Awolowo may very well have seen no reason to be flexible whenever principles are concerned. By principles here we mean those principles that are associated with the regime of mental magnitude and which alone could lead to Awolowo's ideal of a socialist state. It is perfectly in order for him to argue that mere adherence to principles is not inflexibility: that the major cause of indiscipline among the rank and file of Nigerian leaders is just this lack of adherence to principles. This leads to a complete disregard for the rule of law in the Nigerian society. He may, like Plato, think that until kings are philosophers or philosophers are kings, and until leaders of state are ruled by reason as opposed to appetite and negative emotions, there can be no just, egalitarian, and disciplined society. He might then argue that if his opponents feel unable to compromise with him on principles, it is precisely because of their inability to cope with the demands of the regime of mental magnitude. As Awolowo himself has stated somewhere in one of his books, there are those who would regard the prescriptions of mental magnitude as too stringent for leadership. If this is the case, he argues:

> They are welcome to their view; but for the good of the fatherland, such people should steer clear of the affairs of state, and confine their activities to those spheres where their excessive self-indulgence cannot incommode the entire nation, to the point of threatening its very life.[65]

From this point of view he may conclude that only those with like minds, those possessing mental magnitude, can effectively work together for the ultimate purpose of the regime of mental magnitude which Nigerian society and other African societies badly need. I think, therefore, that Awolowo would be right to contend that if his alleged insistence on rational principles of mental magnitude seems to have been the main reason why he had failed to rule Nigeria, then so much the worse for the country. In this way he could argue, like John Stuart Mill,

> It is better to be a human being dissatisfied than a pig satisfied; better to be a Socrates dissatisfied than a fool satisfied. And if the fool, or the pig, are of a different opinion, it is because they only know their own side of the question. The other party to the comparison knows both sides.[66]

And he may, like Socrates, rather drink the hemlock than give up a cherished political philosophy which forbids the causes of those maladies that had made, and are making, efficient governance virtually impossible in Nigeria and many parts of Africa: greed, corruption, nepotism, mediocrity, indulgence in alcoholic beverage, and, the worst of them all, sexual indiscipline at home and abroad.[67]

The rejection of Awolowo's political ideals constitutes a tragic paradox. The same doctrine which in Nigeria is rejected by default is the one that is badly needed to keep the nation on the move. Nigeria witnessed four military coups between 1966 and 1984 for the very reason of indiscipline, greed, corruption, nepotism, lack of direction, and all the negative emotions which Awolowo's mental magnitude was designed to curb. Even the government which rejected Awoism later introduced Awoism through the back door when it frantically called for an ethical revolution early in 1983. But the call for an ethical revolution in Nigerian society failed woefully because the revolution did not start from the top, from the leadership, which in Awolowo's words must be self-disciplined if it is to succeed in disciplining others. Once again, Awoism was confirmed as the better political institution for Nigeria, and perhaps other African countries when, in the face of various malpractices among politicians, and

79

the misdirection of government priorities, the Nigerian army struck on 31 December 1983, the fourth time since independence. With the exception of Tanzania, many African countries have tasted one or other military coup for much the same reasons of greed, corruption, and misgovernment or, if you like, lack of the regime of mental magnitude. The question to be raised, therefore, is this: Is Africa unprepared for or incapable of the regime of mental magnitude which Awolowo sees as the sine qua non for democratic socialism?

If one understands that Awolowo appreciates the developmental efforts of capitalism, and is a great advocate of democracy with spiritual depth, the essential differences between his socialism and capitalism under democratic institutions seem to narrow down to the evils he sees to be associated with the capitalist method of social, political, and economic organization. These evils are greed, exploitation, dehumanization, lovelessness, and total disregard and disrespect for personhood, all of which are the evils associated with an extreme materialistic conception of human beings and the world. That is to say, Awolowo's regime of mental magnitude exhibits itself in capitalistic societies only insofar as such societies have employed their reason through rigorous education to conquer their environments for the purpose of meeting their needs, organizing free and successful elections, maintaining stable economies and political systems, and improving the material lot of their people. But he would contend that capitalism has not used the regime of mental magnitude to improve the general conditions of human beings such as would make them live a fully, happy, and triumphant life, free from exploitation, dehumanization, loss of self esteem, fear, and anxiety about the future of the human race. The same would be true of atheistic communism, irrespective of its social and political ideals. What is needed in both is mental magnitude and spiritual depth, with strict adherence to the golden rule--love thy neighbor as thyself. It is only when this injunction is obeyed that we can boast, in Awolowo's sense, of the regime of mental magnitude.

Some criticism could be made against the role of God in Awolowo's socialist doctrine. It may be argued, for instance, that his view seems to confirm John Mbiti's claim that Africans live in a religious universe. It may also be argued that it is not usual for theoreticians to ascribe an important role to God in the development of scientific, social scientific, and political theories. A philosophical atheist, like Nagel, may argue that a reference to God in science, or socialism as a normative science, is not only unscientific but fundamentally otiose. Nagel has argued, as indeed have many philosophical or scientific atheists, as follows:

Atheism as a philosophical position is directed against any form of theism, and has its origin and basis in a logical analysis of the theistic position

80

and in a comprehensive account of the world believed to be wholly intelligible without the adoption of a theistic position.[68]

On the possible role of God in scientific inquiry, the French mathematician, Pierre Laplace, was once reported to have made a personal presentation of a copy of his book on celestial mechanics to Emperor Napoleon Bonaparte. When Napoleon glanced through and found that nowhere in the book did Laplace make any reference to God, he was quick to ask about the role God played in Laplace's book. Laplace's response was very instructive as it was indicative of the atheist position: "Sir, I have no need (place) for that (sterile) hypothesis," was the reply Laplace was reported to have given Emperor Napoleon. According to Nagel, the dismissal of sterile hypothesis such as the existence of God "characterizes not only the work of Laplace, it is the uniform rule in scientific inquiry. The sterility of the theistic assumption is one of the main burdens of the literature of atheism both ancient and modern."[69]

This view seems to suggest that religion stands in the way of proper scientific inquiries, such as Awolowo tries to do for socialism. But aside from this, the kind of objection raised by Nagel can also be raised against what some people have taken to be a religious foundation of Awolowo's socialism, that is the point that Awolowo's socialism leaves too much to the operation of the universal mind which is equated with God.[70] Thus, those who see atheism as a form of social and political protest against institutionalized religion may object to too much reliance on God in the development of a theistic social and political doctrine by arguing, like Nagel, as follows:

> Atheism has been, in effect, a moral revolution against the undoubted abuses of the secular power exercised by religious leaders and religious institutions. Religious institutions have been havens of obscurantist thought and centers for the dissemination of intolerance. Religious creeds have been used to set limits to free inquiry, to perpetuate inhumane treatment of all the ill and the underprivileged, and to support moral doctrines insensitive to human suffering.[71]

Strong as these arguments may be from the scientific point of view, a little reflection will show that its objections against Awoism can be met.

Marxist Totalitarianism and Capitalism's "Totalitarian" Democracy

Awolowo's socialism can be seen as a rejection of atheistic communism. It is also a rejection of atheistic capitalism. While the former believes in the one-party totalitarian state and the use of force, oppression, and curtailment of freedom to achieve a socialist ideal, the latter, although operating under democratic governments, is hardly better. Government is usually that of the elite, the few wealthy and powerful members of society who defend their economic interests through various means, including the killing of their fellow men. While the Marxist totalitarian regime is abhorrent, totalitarianism finds its way into democratic institutions through capitalism, an extreme form of materialism.

In a brief address to the people of Gettysburg, the President of the United States of America, Abraham Lincoln, had this prayer: "that the government of the people, by the people, for the people shall not perish from the earth."[72] A good prayer it was, in what has been the most accepted definition of democracy. But it is only a definition. The word "people" as it appears in Lincoln's statement does not mean the same thing in each of its occurrences, as it does not refer to the same kind of people. The best example to substantiate our position is in George Orwell's 1984.

In his discussion on what he calls "the theory of and practice of oligarchical collectivism" George Orwell makes it clear by implication that a purely democratic government is an illusion.[73] It seems reasonable to assume that, for the purpose of human convenience, a great number of people will be prepared to accept the views and authority of a few as representative of their own views and authority. But the danger of this democratic characteristic is that it makes totalitarian government easy once the few have been given the instrument of power and persuasion. The underlying factor here is the human greed and appetite for power, wealth, and domination. Thus Orwell points out that the selected group of people to whom power is given to govern the society is known as the Party. "Individually no member of the Party owns anything, but collectively the Party owns everything and disposes of the products as it thinks." At the very top of the pyramid of Orwellian society is Big Brother from whose leadership and inspiration every success, achievement, victory, scientific discovery, all knowledge and human wisdom, happiness, and virtue are derived. "Big Brother is the guise in which the Party chooses to exhibit itself to the world. His function is to act as a focusing point for love, fear, and reverence."[74]

Big Brother exists in both democratic and non-democratic societies. In him we find the faces of power by which human beings are deceived, exploited, dehumanized, and effectively controlled by means of what Orwell calls "doublethink." It means "the power of holding two contradictory beliefs in one's mind simultaneously, and accepting both of them." Thus, doublethink, falsification of facts or history, and mind control are needed to safeguard the interests and the infallibility of the Party as well as the omnipotence of Big Brother. Paradoxically, the day-to-day falsification of history is the duty of the Ministry of Truth! It is a system by which the Ministry of Peace concerns itself with war, and the Ministry of Plenty with starvation. Well may Orwell proclaim doublethink as "a vast system of mental cheating."[75] Therefore, doublethink is an irrational system of belief--belief in contradictory terms and propositions only insofar as such beliefs are to the interests of the Party and Big Brother. In a democratic institution where doublethink is covertly or overtly present, there exists Big Brother and one form of totalitarianism or the other. This is not restricted to any region of the world, or to powers. There exist Big Brothers in Africa.

The road to a totalitarian regime differs only in methods, not in kinds. While the one achieves it by force or a non-democratic one-party system, the other achieves it through a system whereby once in every four or five years at the general elections, people are made to believe that they are governing themselves. As soon as a few people come to power, they own everything and dispose of the products as they think, to the extent that a government of the people by the people and for the people becomes a government of the people by the elite, for the elite. What is more, in order to hold on to power, every system of mental cheating is employed, especially through the mass media. In the underdeveloped countries mass rigging of elections, persecution, killing of political opponents, bribery, and corruption are the known added attractions. With all these, democracy takes up the garb of totalitarian democracy, different only in certain degrees from Marxist totalitarianism and dictatorship.

The global conflict of the First and Second World Wars, the Russian Revolution, Nazism, and its bestiality, the aftermath of wars that led to dictatorships and the development of destructive weapons all over the world clearly indicate the non-rational element of human nature, its insatiable appetite for power, material wealth, and desire for territorial ambition, all of which threaten the survival of the human race and civilization. Yet all of these have been aided by scientific intelligence, the use of reason totally devoid of spiritual depth. If Nagel's arguments were valid, we would have to blame religion rather than science and technology for social and political injustice, greed, exploitation, man's inhumanity to man, erosion of privacy through electronic surveillance, ideological warfare, and the daily threat of nuclear holocaust.

The issues raised in Awolowo's reference to God and the Christian injunction introduces a moral dimension to the system of social and political organization. Even if there were no God, it would still be difficult to find anything morally objectionable in the biblical golden rule "love thy neighbor as thyself." Even before I read Awolowo's People's Republic I had argued that this biblical injunction appeared as the most humane principle of human association we can get, although it appears that this injunction can only be approximated, as no mortal seems capable of obeying it at all times.[76] But the injunction is of extreme importance for the improvement of social and political organizations, whether at the local or international level. Auden has said "we must love one another or die."[77] This is not the spirit of science but of religion, and particularly of the golden rule. If people love others as they love themselves, do unto others as they wish other people to do unto them, society and the world would be a better place in which to live. Greed, selfishness, doublethink, oppression, apartheid, racism, nepotism, exploitation, terrorism, cold war mentality, and all the negative emotions would vanish. Again, there is nothing one can say against the development of this religious spirit as an attitude of mind for it alone seems capable of guaranteeing a much needed spirit of love, brotherhood, and selflessness.

In fact, the religious spirit of love thy neighbor ought to be the supreme principle of all systems of morality behind social and political systems. As some religious scholars have observed, the portable phrase "love thy neighbor as thyself" points to the individual as well as to the collective aspects of manhood and brotherhood.

> It shows us that the real need is to bring into our lives the ideas and ideals of the Christ message of love so that we can develop unity of purpose, carve out noble life motives, and see them manifest in thoughtful considerations, genuine sincerity and pure selflessness.[78]

In order to do this, we must be able to relinquish our interests and desires in the love of others. Therefore, in no way must religion be seen as anti-science unless, of course, science exists for evil purposes. Perhaps, in the end, we shall have to come to terms with a hard fact as recognized by Langdon Gilkey, a brilliant commentator on religion and modern scientific culture:

> Modern culture in the development of its science and technology has not made religion irrelevant. It has made religious understanding and the religious spirit more necessary than ever if we are to be human--and even if we are to survive.[79]

From all this the religious dimension in Awolowo's socialism is intuitively plausible. On this score we might say, as did Voltaire, than even if there were no God, we would have to create one to help humanity tread the path of love, selflessness, brotherhood, and social justice. If human beings, as political animals, must live peacefully and happily together in a society, both the leaders and their followers must be rendered good lest communal living prove a curse. This seems to show beyond any shadow of doubt how eager Awolowo is to show that socialism is not "anti-God, anti-Christ, and anti-Mohammed," and to find every device and policy that would prove human beings, either in the

84

assessment of the Psalmist as only "a little lower than God," or in the view of the atheist and agnostic as just homo sapiens, as capable of governing themselves in freedom and virtue. That is the quintessence of Awolowo's socialism which, from the point of view of his own discussion, raises the question as to whether or not there can be an African socialism, a socialism which takes its roots from Africa or any region of the world.

In anticipation of an objection that his socialist ideal may be far from being realizable, Awolowo argues as follows:

> To those who will rejoin that these ideals are too lofty for human achievement, we quote the eternal words of Jesus Christ who never enjoins man to impossibilities. Says he: "You shall know the truth and the truth will set you free."[80]

By the same token Awolowo feels that some day we all will know the truth and enter into the regime of mental magnitude where love reigns supreme. From this concluding remark Awolowo seems to have given the impression that his brand of socialism may not be for the present, but for future Africans. But then, if Awolowo's brand of socialism were to succeed in Nigeria and other parts of Africa in the distant future, it would not be called Marxist socialism (Marxism) but Awolowo's socialism (Awoism). If this were to happen, African socialism, precisely what Awolowo does not believe in, would differentiate Awoism from Marxism. Surely socialism as a political and philosophical doctrine does not have the kind of universal validity one finds in logic, mathematics, and the advanced sciences.

As with its philosophy, as we pointed out from Russell earlier in this work, a people's system of social and political organization may be determined by the circumstances of their lives. And as Awolowo has argued:

> Whether we acknowledge it or not, the fact is incontestable that our own tendencies and habits, plus those we have inherited from our ancestors are . . . decisively influenced by our environment. The food we eat, the clothes we wear, the style of our buildings, our temper, prejudices, and affections, our mode of thinking, the language we speak . . . are mainly and decidedly the results of heredity and environments.[81]

But it is precisely from our temperament, the language we speak, and our mode of thinking, all of which are relevant to our cultures, that African philosophical, social, and political thoughts and ideas develop. If, as we have argued, there is an African philosophy, a Chinese philosophy, a British philosophy, and an American philosophy, there can be an African socialism, a Chinese socialism, a British, or an American socialism. There is, in fact, the British socialism associated with the Labour Party in Great Britain, with a history of strong association traced to the philosophical writings of John Stuart Mill and the Fabian Society.[82]   Although Awolowo has a strong tie with the British Labour Party, British socialism is not Awoism.   Rather Awoism would be seen as African socialism precisely because, in Awolowo's view, socialism as portrayed in the West by Marx and Engels is not relevant to contemporary Black Africa.[83]   And if Awoism is a kind of socialism that is relevant to African circumstances, what can it be but an African socialism?

Now, if we look back to the question as to whether or not there is African philosophy, it shall be seen that our position in the present work has been in the positive. And with regard to the other question as to who is an African philosopher, we were led by the twist of argument to see that many contemporary Africans, including Awolowo, Nyerere, Senghor, and Nkrumah, are African philosophers.   Awoism, no doubt is an authentic social and political philosophy, and Awolowo's theory of the regime of mental magnitude has been rightly seen by some writers as a valid and necessary message of self-discipline and reason.   And when seen as "the totality of the theories or doctrines associated with the ideas and teachings of Chief Obafemi Awolowo of Nigeria" Awoism appears to me an African socialism, in much the same way that our examples of Yoruba or Akan philosophical thoughts and systems of traditional medicine were seen as African philosophy and African traditional medicine respectively.[84]   Accordingly, we conclude that Awolowo's contention that there is no African socialism may very well be mistaken, for such a contention is by no means uncontroversial.

Chapter 5

AFRICAN TRADITIONAL MEDICINE:

PRINCIPLES AND PRACTICE

Apart from Professor Abimbola's writings and my discussion on Ifa as a repository of knowledge in Yoruba thought, it has been stated that the works of Zahan on Bambaras and Thomas on Dolas have also afforded us an opportunity of forming some ideas of the theory of knowledge among peoples who possess neither writing nor machines. Eberhardt reports of three degrees of knowledge among the Bantus: superficial knowledge, that of the bush or the "third world"; the initiation form of knowledge of the second or intermediary world; and the weighty or profound knowledge of the first world. Zahan has also stated that the Bambaras possess a systematic view of the world which endeavors to provide an explanation for both the macrocosm and the microcosm by devising, without writing, a number of graphic signs of which some scholars have identified 266. Thus arithmetic and numbers are used to define numerically the constitution of matter and of the world in general, all of this giving rise to a synthesized overall picture of the cosmos and the principal natural phenomena.[1]

In the field of biology the Bambaras know the principal organs, some of which are regarded, as they were in ancient Greece, as the seat of the moral and intellectual faculties. The will is taken to be in the kidneys, the power of judgment in the liver, fear in the bladder, and courage in the heart. They also have an explanation of speech that implies the interaction of seventeen organs, with the ritual number of human teeth as forty-four which have an effect of enlightening the spoken word. Therefore, all the devices for polishing the teeth, toothbrushes, chewing sticks, and toothpicks are all means of speech conditioning.[2]

From Abimbola's writings, oral evidence, and my analysis of Ifa as a repository of knowledge, an epistemological basis for African traditional medicine becomes readily available among the Yoruba of Nigeria. We hope to show that African traditional medical assertions occur already in an epistemologically consti-tuted universe. As I have said before:

Ifa has been called by some people one of the angels of God. It is a Deity, identified with orunmila, the owner or possessor of wisdom and knowledge. Through Ifa, orunmila brought wisdom and knowledge into this world. Such knowledge consists of several branches: science of nature (physics), animals (biology/zoology), plants (botany), medicinal plants (herbalism), oral incantations (ofo and ase/afose), and all the sciences associated with healing diseases (medicine).[3]

The knowledge of healing derived from Ifa is known as complete cure (iwosan).[4] Such cure is possible because Ifa knows the origins of diseases and the various names by which they are called. It also knows the leaves, herbs, roots, and animal substances associated with the cure of all diseases. In the opinion of an internationally well-known Ifa priest, Ifa is the controller of language, culture, philosophy, and religions among the Yoruba. In short, its adherents believe Ifa to know the causes of things and events, the names and nature of things, as well as their origins and chemical compositions.[5] Thus Ifa is, like Laplace's omniscient intelligence, or the Magus of Bombastus Paracelsus, the wise man to whom nature has taught her secrets. The Magus has a command of the forces of nature, and knows the signs which reveal her powers.[6] It is from the immense reservoir of Ifa's knowledge that various branches of knowledge emerge, including knowledge of traditional medicine. This is the knowledge of the nature and use of animal and plant substances, incantations, and authority or power of words for medicinal purposes (both preventive/protective and curative medicine) as a way of prolonging lives in the face of illness, diseases, and evil forces on earth.

Ifa is said to know the diseases in the world, the names by which they are called, and the power (ase) of each of these diseases. The treatment of a disease is the application of what the disease is forbidden to take, at whose sight the disease must disappear. Thus Ifa is regarded among the Yoruba as the path-finder (atoka) of medicine and healing, and the source of our knowledge of herbal and metaphysical medicine (divination and oral incantations), the kind usually referred to in the West as magical science.[7]

Osanyin, the father of traditional healers (onisegun) was Ifa's slave to whom he taught herbalism (egbogi), the use of natural herbs, leaves, and roots for medicinal purposes. The totality of traditional or native medicine in Yorubaland is Ifa divination, oral incantations, and herbal medicine.[8] While traditional healers often specialize in either of these aspects of medicine as Ifa priests/babalawo (diviner) or onisegun (herbalist), some combine both. Since Ifa knows the nature of all illness and diseases, and the medicinal plants or animal substances that could bring complete cure, a medical herbalist (onisegun) may still consult Ifa in order to get the appropriate remedy. He can do this by learning Ifa or by consulting with an Ifa priest (babalawo) whose specialty is Ifa divination. An Ifa priest may also acquire a knowledge of herbal medicine if he has a flare for it. However, a combination of both knowledge of Ifa and herbal medicine does help a traditional healer in effecting complete and permanent cure of some illness, whether organic or functional, including those beyond the competence of modern physicians.[9]

What is meant by knowledge of medicine and good medical practice in the Yoruba traditional setting is to know the root of diseases or illness (gbongbo arun) as well as the most effective remedies by the use of particular herbs, leaves, roots, and animal substances as they are revealed to Osanyin by Ifa. Herbalism may thus be combined with Ifa divination and oral incantations. Hence, whenever a patient is given an effective treatment to such an extent that the illness never recurs (such as a mental disorder or a malignant disease) the Yoruba say that the patient has been treated from the very root of the illness or disease (wa egbo dakun fun alaisan or wa arun t'egbo t'egbo).

One remarkable feature in the Yoruba traditional medicine is the healers' peculiar understanding of the nature of things, an understanding derived from Orunmila's knowledge of the names and nature of things. Thus, while the traditional healer, a medical herbalist, knows the correct medicinal plant needed to treat a headache, and prepares the drug either in the form of a potion or powder, he could go further to make his treatment more effective by consultation with Ifa. The Ifa priest, using his knowledge of the names and nature of things gained from Ifa, may make a symbol of treatment on iyerosun, a light, yellowish, powdered substance extracted from the irosun tree, call the headache by its original name, disclose its nature and secret and where it got its power to trouble human beings. The power of the babalawo or onisegun derives from his understanding of the nature and secret of headaches. Considering the headache as an example, Ifa knows its weakness and the things which, when given to the patient, kill the headache. The complete cure of headache is therefore the application of what it is forbidden to take. The calling of its original but secret name and the application of that thing which it is forbidden to take (usually in the form of medicine) cause the headache to vanish from the affected person. In other circumstances the headache may not respond to treatment if it is handled by a quack traditional healer who has no know-ledge of the secret or power of headache and the appropriate medicine for its treatment. The same pattern of knowledge of the nature and secrets of things is used by all good and reputable traditional healers, the babalawo (diviners) and onisegun (herbalist) or diviner-herbalist who are well trained in their respective fields.

Quite apart from their familiarity with what some scholars call native botany and a knowledge of plant poisons and antidotes, as well as dosage, the African traditional doctors are quite familiar with the principal organs of the human body.[10] Each organ of the body has its part to play in the human biological system. For example, the Yoruba note that the emi (soul) is seen as a living or vital force of the human body. Abimbola has argued that the physical counterpart of emi as a living or vital force of the body is the human heart.[11] This belief is probably due to the important role the heart plays in the human body. But from oral evidence, my father, a traditional healer of great repute, suggested to me that the emi (soul) as a living or vital force in the human body has its physical counterpart not only in the human heart but also in the human blood. For this reason the heart and blood are seen as inseparable companions in the services of the soul (emi), the thing that gives life to the body.[12] Any question as to which of these two is more important than the other as the physical counterpart and agent of the soul would be like the chicken and egg question. Surely a heart without a supply of blood is no heart, and blood without the heart as its pumping machine is useless. That is to say, a heart without blood is lifeless; blood without heart is idle and useless. Therefore, the special recognition given to the importance of both the heart and blood in the human body, where both are regarded as agents of the soul in the physical plane, has a great influence on the African theory and practice of traditional medicine, in which blood in particular is regarded as the ultimate beneficiary of all medications. The importance of the blood as representing the soul in the physical plane is suggested by the saying: the life of a man is in his bloodstream.[13]

Although the Yoruba believe essentially in the dualism of the body and soul, they hold a tripartite conception of a person. Quite apart from the two main elements of a person, ara (body) and emi (soul), they believe in a third element known as ori (inner head). We have also noticed that, as a metaphysical head, ori is the bearer of human destiny. For this reason, part of the treatment of an illness may involve sacrifice to one's ori as part of the human effort to avert a possible premature death. Since the Yoruba believe in the immorality of the soul, the spirits of the dead, particularly of ancestors, are sometimes appeased through propitiation or sacrifice to aid the course of a particular treatment. The majority of human suffering and sudden death are attributed to evil forces--wilful maleficia--who are said to do bodily harm to their fellow human beings.[14] As a way of combating these evil forces, the Yoruba believe in appeasing the evil spirits through sacrifices (ebo/etutu). In some cases protective counter-medicine like madarikan or amulets of various shapes are used. In other cases, a direct attack by the means of a curse (epe) or incantation (ofo/ase/afose) is used with good results. However, as far as the diseases of the body and all organic illnesses are concerned, the blood is regarded as the most essential organ to the body. It is the thing that actually

feeds the body and makes it grow. Hence the b
treated with good care. Since the blood is also
beneficiary of the food we eat, there is seen an imp...
tion between food and blood, on the one hand, and betwee...
blood and good health on the other. Thus good food is seen as a
kind of preventive medicine, for it keeps the blood pure, rich,
and healthy. Healthy blood in this sense is taken as synonymous
with health of the body.

Now, when we look back and consider how the knowledge of
traditional medicine has grown in some parts of Africa, we see
that successive cultural and philosophical systems have played
positive roles in it. But the general opinion is that African
traditional medicine has led to no discovery in the medical
sciences. This needs an explanation. African traditional
medicine had existed before the introduction of Western medicine,
and suited the needs of African cultures. Therefore theories and
practices of medicine have a cultural dimension. Its systems and
policy must have taken root from the beliefs and cultures of the
African people. From this point of view the concepts of illness,
disease, diagnosis, treatment, life, and death must also have a
cultural dimension. That a people's belief, faith, and religion
have much to do with the acceptance of the efficacy or otherwise
of any particular system of medicine need no philosophical
justification. The main reason why traditional medicine has led
to no discovery in the medical sciences is that while the genuine
traditional healers use purely African remedies that are often
wonderfully effective, they usually do not reveal their secrets
to anyone except their children or immediate relatives, so that
one finds no textbook on traditional medicine. Therefore, no
meaningful research can be carried out for the purpose of
discovery and growth in medical knowledge.

African medicine has produced a number of genuine healers
who have learned through years of study and experience to
recognize the symptoms of sickness and disease and to apply
effective remedies through knowledge of medicinal herbs and other
traditional means. The secretive nature of their knowledge,
however, has made the principle and practice of traditional
medicine inaccessible to curious minds. As a result, all the
knowledge of traditional medicine dies with traditional healers.
This situation is so true that a writer has said of them as was
said about the "Griots": "The death of a genuine healer is
tantamount to the loss of a library."15 The current dispute
between some Western-trained physicians and traditional healers
in Nigeria rests on this secretive nature of African traditional
medicine and traditional healing.

However, the dispute is more on the side of one aspect of
traditional medicine usually referred to as "magical" than on the
herbal medicine which is always open to empirical investigation
and rigorous tests. In this regard, relying purely on scientific
proof, modern doctors tend to reject the activities of diviners
(babalawo) and the oral healing provided by the herbalist

91

onisegun) with the aid of incantation (ofo) and ase or afose--
the last two being regarded as the most powerful aspects of
traditional medicine.[16] Unfortunately, these aspects of traditional medicine are the ones that are not understood by Western
doctors because they are not open to empirical investigation.
This is the situation that led to the common distinctions between
scientific and nonscientific ways of knowing or between
scientific and magical ways of knowing.[17] For instance, the
phenomena of a physical or medical laboratory, however new or
unprecedented, are very far from having the character of miracles
in the sense of magical or supernatural events. It is true that
while herbal medicine is open to laboratory tests, traditional
oral medicine, such as ofo or afose, is not, and its methodology
is quite unknown to empirically-minded scientists. Hence this
important aspect of African traditional medicine is often characterized as "magical." It has been suggested, however, that both
magic and science can be perceived as complimentary when problems
of health and ill-health are being considered. It is from this
point of view that modern doctors should look more carefully at
the claims of traditional healers who employ ofo and afose for
both preventive/protective and curative purposes, and see what
integration could be made between herbal and oral medicines.
Although a scientist may argue that it is pretty obvious that
there are good scientific reasons for preferring science to
magic, presumably, there would be good magical reasons for
preferring magic to science in certain cases and circumstances.[18]

African traditional medicinal herbs and medical practice
imply the existence of the nature, classes, or kinds of causes
and effects which empirical investigation brings to light through
observation and inductive method, followed by tests and verifica-
tion from specific deductions. On the other hand, oral medicine
implies the existence of latent kinds of natures which mystical
research contemplates as magical or supernatural. Thus oral
medicine (Ifa divination and oral incantations) has its metaphysical foundations in the nature of things. In addition, oral
medicine demonstrates to the fullest the power of words and the
relation of these words to natural phenomena such as the herbs
and animal substances which are used as ingredients for such
traditional medicine as ase or afose. Here we may note that ase
or afose consists of natural ingredients chemically diluted into
powdered form and put on the tip of the tongue before incantations are said. The substances are either from animals or plants
or both.

A more powerful ase is packed in the horn of particular
animals with particular powers attached to their natures.
Typical examples are the horns of antelope, deer, ram, or
buffalo. The plant and animal substances used for ase are such
that their natures must be known, as the words spoken in incantations must reflect the natures of these substances as well as
their possible effects on human beings, disease, and illness. As
a typical example, part of the medicinal ingredients used in ase

92

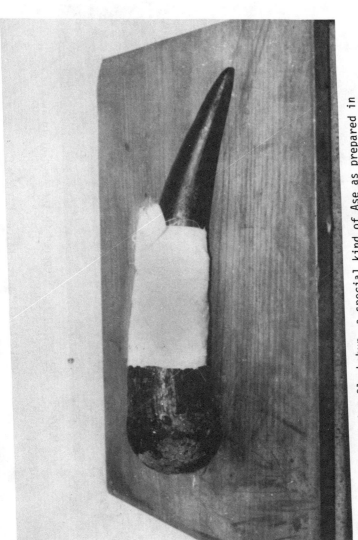

Figure 1. Olugbohun, a special kind of Ase as prepared in
the horn of a bull and partially wrapped with
white cloth. It is one of the most powerful.

(Courtesy of Daniel Kayode Makinde)

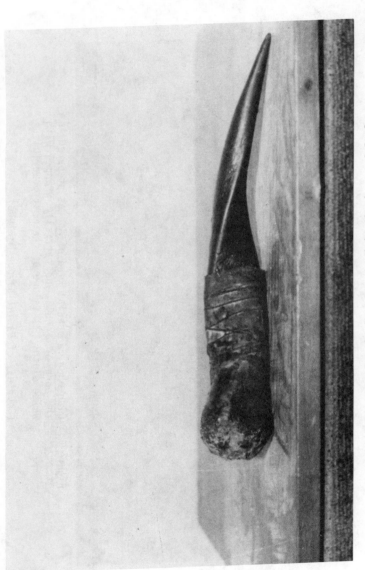

Figure 2. Ase Iwo Igala, prepared in the horn of Igala, a species of deer.

(Courtesy of Daniel Kayode Makinde)

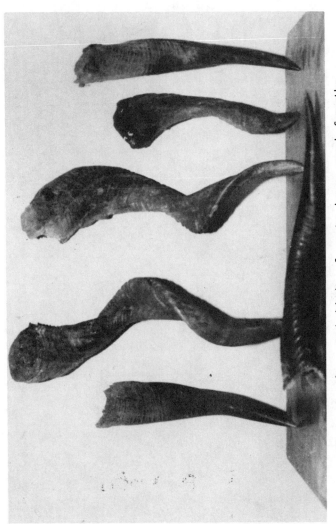

Figure 3. Assorted kinds of empty horns used for the preparation of Ase of various kinds. They range from the horns of antelope to those of buffalos.

(Courtesy of Daniel Kayode Makinde)

consists of the tail of a big rat (iru okete) and the whole of a cameleon (alegemo) together with plant substances of particular kinds. One significant aspect of the big rat is the power given to it by nature. In digging a hole for its habitation, it does not seem to exert any energy. It simply uses its tail as if it commands the earth to open up. As for the cameleon, it has the power to change its colors as it wishes.[19] After these animal substances are mixed together with other plants with peculiar powers, the mixture is put in the horn and neatly tied in black and white cotton to prevent them from falling out.

In its use the traditional healer touches the tip of the medicinal horn with his tongue, holds on to the horn with his left or right hand and says whatever he wants to happen. Then he invokes the mysterious power of each of the animal or plant substances included in the preparation of the ase. The combination of the wonderful powers inherent in the nature of these plants and animals, and the words that relate such powers to the human understanding of their possible effects on things or situations in the outside world, gives ase its peculiar authority and power. For instance, in using the ase, the herbalist says the following incantations among others: "Aba ti alagemo ba ti da ni orisa oke ngba; Oro ti okete ba ti ba ile so ni ile ngba," (whatever suggestion of color a cameleon makes is always accepted by the heavens or the deity; whatever the big rat tells the earth it accepts without argument or complaints; therefore, whatever I say must come to pass).[20] And when a good ase is used in this way by a competent herbalist, his words, as expressed in incantations, do come to pass! That is why in Yoruba society a man with ase is a powerful person and he is respected for that. And the words of ase can only be overturned by means of a counter-medicine.

A counter-medicine is usually protective in nature. It could be in form of igbere (native injection), or oruka (medicinal ring). But the most popular of all is madarikan, literally meaning "do not knock your head against mine." It is one of the most powerful preventive and protective traditional medicines. A very powerful kind of madarikan consists of the bark of the mysterious asorin tree. Wherever it grows, no other tree grows near it. It kills other trees if their roots touch its own. Even the traditional healers or herbalists rarely touch it. It is a very difficult tree. The nature of asorin has given the herbalists an insight into its value: Igi to a ba fi egbo kan asorin kiku ni ku (any tree that touches the root of asorin with its own root must die). In this way the herbalist uses egbo asorin (the root of asorin) as an ingredient in preventive/ protective medicine. When it is so used, it is known as ajesara (power given to the body to ward off evil spells, a counter-medicine). Nowadays, madarikan is sought by any person who aspires to prominence in society. The wealthier you are the better quality of madarikan you get, as there are various kinds of them.

The need for counter-medicine is clearly shown in the possible effects of a certain destructive traditional medicine (ogun) used by people in place of the guns of developed countries. Justice Adewale Thompson gives a few examples under the title "Wonders of Our Land." For instance, owo, already mentioned above, is a charm which has the effect of warding off evil.[21] If kept on the body no mishap will occur to that person. If kept in a car, the car will escape accidents. Thompson supported his view with a most extraordinary experience he had about owo when he defended a young man accused of petty theft and got him discharged and acquitted. The accused father, a night guard, had given his son the charm. After the case, the parents came to thank him, and gave him "two small emblems, each of which looked like a piece of thread woven round an ordinary piece of stick." These were the owo charms which the man asked him to keep, one in his car and the other at home, with the assurance that no evil would come near the emblems. Although at first Thompson regarded such things as too fantastic to be believed, his experiences and those of his wife at home and on the road bore testimony to the wonderful affects of these charms.

Another brand of owo is used for the apprehension of robbers. There are many versions of this charm. A popular one known among the Yoruba is made of two brooms magnetized with magical preparations and kept either at the entrance to a home or in a very conspicuous place inside the home. My father called it imole (thief catcher). Like the computer, it sometimes replaces human labor as a security man. As soon as the robber enters the premises, he picks up the brooms and begins to sweep the whole place until the owner of the house sees him. Unless the effect of the spell is removed, the thief continues to sweep. There was the case of a pharmacist at Ibadan, Nigeria, whose shop was invaded at night by a gang of three robbers, two men and a woman. The three robbers forgot their mission and instead started to sweep the floor until the following morning when they were apprehended. It made a sensational news item in the newspapers.

Among villagers, adultery is frowned upon. In order to prevent a wife from being sexually assaulted, the husband consults with a herbalist or babalawo who gives him the dangerous charm known as magun (literally meaning "do not climb" or "mount"). The man gives it to his wife to drink.[22] From then on any other man who has intercourse with the wife finds himself somersaulting, just at the point of orgasm, until he gives up. The more popular kind of magun is not administered by mouth. It is put on a thread and secretly left on the ground so that the woman walks over it without the slightest notice of it. It is then removed to prevent other women from catching the dangerous magun charm, which is considered as the eradicator of adultery in African village communities. However, magun has antidotes, some of which have been suggested by Thompson. Some of these are in the form of anti-magun charms, and others are provided as treatments. I had myself witnessed from my late father the treatment of men caught by magun. Although some Western-trained Nigerian medical doctors had debated the reality of this dreaded charm, none of them had the courage to provide themselves for experiments by which they might have confirmed or disconfirmed magun's efficacy. It is, however, instructive to point out that herbalists do sometimes test their magun on animals. In order to put an end to the controversy surrounding the reality and efficacy of magun and other dreaded charms in Yoruba traditional medicine, some courageous Western-trained medical doctors could, if they remained unconvinced by the power of their own traditional medicine, offer themselves for experiments in the use of magun, ase, afose, and epe.

Apart from magun, there are some native medicines which can be used for evil purposes. Thompson gives a few of these as follows:

1. Apepa: A charm by which a person's higher self could be evoked and banished from physical existence, thereby causing instantaneous death of the victim.
2. Olugbohun: A powerful charm believed to be a representation of the echo and which is reputed to act as a catalytic force to the power of words which the ancient mystics have laboured so hard to find. This can be used for good or evil. When used for good, it can, like afose, be used for prayers. Its evil use can take the form of epe (curse or imprecation) which is another version of the charm of the spoken words. In the Yoruba tradition (according to my late father) the people of Egba are said to be the owners of ase, and the Ijebus the owners of epe. Both the Egbas and Ijebus are in Ogun State of Nigeria.
3. Gbetugbetu: Like ase, afose, olugbohun/epe, this charm is also related to the power of the spoken word. In Yoruba tradition the owners of Gbetugbetu are the Ekitis in Ondo State of Nigeria. Gbetugbetu is used during a general crisis such as war or when a person is face to face with danger. When used, the spoken words are like a powerful command that must be obeyed unless, of course, the person to whom the words are targeted is powerful enough to neutralize the possible effect of the power of spoken words derived from the use of this extremely powerful charm. The person to whom the spoken words are targeted is usually in sight. He does everything he is asked to do. If the victim is in possession of a gun and wants to shoot, he is asked to drop it. He may also be asked to remove his clothes, carry them on his head, and run around the place or even into the bush from where he may never return.
4. Apeta: A less virulent version of Apepa. According to Thompson, the victim's higher self is summoned into a

reflective substance and shot with a gun. If the wound is serious enough to cause death, the physical man dies from the gun shot. Apeta presents a problem to modern doctors because, although symptoms of gun shots are seen, there are no visible marks to assist diagnosis.[23]

In a speech delivered by Mike Omoleye at the ceremony honoring the publication of Thompson's book, Omoleye supported Thompson's writing with some examples from well-known Europeans who once lived in Africa, notably in Nigeria and Ghana, whose experience forced them to believe in the power of African traditional medicine. For instance, Rev. J. Buckle Wood, a missionary in the Yorubaland of Nigeria for forty years, was said to have confirmed the reality of the power of African medicine. But a much more interesting revelation came from Dr. Stephen S. Farrow whose doctoral dissertation at the University of Edinburgh (1924) was said to have been written on a related topic. Farrow's subsequent personal experience as a missionary in West Africa led him to accept the reality of the power of African medicine. He even made a strong case for Apeta in his exposition of what he described as he power of African science. He wrote as follows:

> But the strongest and most terrible exercise of this mysterious power is experienced in the dreaded practice of APETA, i.e., "invocation by shooting." A person desiring to kill any one against whom he has a grudge makes a mud image of his intended victim, and at night sets it up, calls the name of the foe three times, and then shoots at the figure with a miniature bow and arrow. At the instant, the victim feels a sharp pain in the region of his body which, in the figure representing him, he has been struck by the arrow. A wound or abscess quickly develops, from which he soon dies, unless he recognises the nature of the injury and can apply a counter charm, a more powerful "medicine" (ogun) than that which has injured him.[24]

It is interesting to know that Farrow himself was a victim of African medicine in spite of his thinking that as a European he was unlikely to be affected by what he called the black man's magic. His experience showed that when he was caught by the black man's power, he found no solution to his mysterious ailment in European hospitals. He was, however, finally cured of his illness by a Ghanian herbalist whom he called by the name of Uncle Teffey. Farrow wanted to share his African experience with others, and, in particular, to show other Europeans that "ignorance of a subject nearly always gives rise to massive disbelief about it." His veil of ignorance having been torn into shreds by his own bitter experience, Farrow could very well accord a great respect for the African "science" or medicine which, as he noticed, could sometimes cause much "indescribable heartbreaks" by causing accidents, producing illness or even bringing about the death of one's enemies. The mystery surrounding this kind of science, one might say, is that it is not in accord with the Western scientific notion of cause, that there can be no cause at a distance. Apeta and other kinds of African medicine either violates this causal law or refutes it. Well could Farrow admit openly that "the evidence for these cases (killing without contact) is so strong that it is difficult to disbelieve them." He also confessed that although he was skeptical at first, he was later "compelled by the testimony of unimpeachable witnesses to modify his views."[25]

Philosophically speaking, we might consider the above as examples of African traditional medicine, many of which make use of the power of the meaning of words. Although not in the medicinal sense, similar trends can be found in the Western world. It is not uncommon nowadays to see various scholars in America specializing in the application of semantics to various problems. I am referring here specifically to the school of general semantics, with its headquarters in Chicago.[26] This school specializes in the application of semantics to various problems, including what language can do to people. In the context of African traditional medicine, the power of words, meaning, and bearing on human happiness, misery, sanity, and insanity, is enormous. The enormous power of the use of words in various incantations and charms derive from the herbalists' or babalawos' peculiar understanding of the nature of things to which these words are related in space and time. In addition to being uttered, some words are chanted while a medicinal preparation is put in the mouth, as in the case of ase or gbetugbetu.

As the examples quoted from some Europeans' recognition of the power of African medicine have shown, African doctors are urged to make careful investigations of this force to see how it could be improved upon, and its system used for constructive rather than destructive purposes. It could be argued that if the advanced countries of the world have their own killing devices such as the guns, Africans could also develop this destructive aspect of traditional medicine as African science or power, to a meaningful stage where it could be used to protect themselves. But since the use of this kind of power demands the knowledge of an enemy's real name and other important details, it may not serve a ready purpose for the destruction of foreign enemies. It could be targeted, however, against particular individuals. The pity of it is that with Western civilization alternatives are being found to invocation by shooting in direct shooting with modern weapons. As for the preventive/protective aspect of African traditional medicine, there is need for modernization. To the Western mind which believes only in Western science and medicine, science is purely physical and empirical. It is for this reason that the Western mind finds African medicine, with all its tested powers, either impossible or just too fantastic to believe. As Thompson says:

> The African is a victim of misrepresentation and obloquy in the hands of foreign travellers not because he has nothing to offer but because, in most cases, he is careless in the manner of presentation of his culture and tradition.[27]

Civilization or modernization, he argues, "is not an exclusive preserve of Europe or America."[28] Rather, it is in the minds of people, whether black or while. As seen by Thompson, the problem is that it appears that the majority of people seem to ignore the modern system of thinking. But another problem is that the majority of African intellectuals seem to ignore the valuable aspects of their traditional systems of thought, belief, and medicine. The majority of Africans have faith in the power and efficacy of traditional medicine but some people in urban areas patronize Western medicine. Because of its usefulness to the society, traditional medicine should be modernized along Western lines through integration. In this case, experts in Western medicine and those in African traditional medicine would become knowledgeable in both systems through medical researches in integrated medicine.

Among the Yoruba of Nigeria, ase is used for complicated delivery, and I have myself witnessed several times its use in breech and stillborn deliveries that otherwise would have been delivered through Caesarian section.[29] But it can also be used for evil purposes, such as destruction of a person either physically or mentally. Since most of the mental illness in Yoruba society is attributed to some evil forces including the evil use of ase by wicked herbalists, nearly all the treatment of mental disorders starts with counter-medicine, especially ayajo or asisan, as a means of removing the possible causes of the illness. It is very common to see mental patients referred to traditional healers for afrotherapy after modern psychiatrists had failed to give proper diagnosis and reliable treatment.[30] Thus, Professor Lambo who agrees with Ackerkrecht that medicine is more clearly a function of the culture than of environmental conditions, rose to fame through his effective treatment of mental patients with both the traditional and modern methods of psychiatry.[31] Together with his colleague, Dr. Tigani E. Mali, he recognized the part played by indigenous psychotherapeutic approach to the total management of patients without any lowering of the standard of medical practice. And he has demonstrated this effectively in his hospital, the Aro Mental Hospital at Abeokuta in Nigeria.

In spite of its high technological and material advancement, Lambo observes that modern medicine has yet to satisfy the basic metaphysical cravings and social needs of many individuals, irrespective of their levels of sophistication.

It is therefore my conclusion, after a long and determined attempt to appraise both indigenous and modern medicines within the African context, that both are not mutually exclusive. In fact, total exclusion of the theories inherent in indigenous medicine from Western medicine would seem to impede the total acceptance of Western medicine by African societies as a meaningful substitute to beliefs and practices which have long satisfied certain basic human needs.[32]

As he further argues, traditional African societies may lack a truly scientific notion of cause and effect. Their traditional ideas of cause are not the result of flagrant theories and fantasies with no basis in reality. Although these ideas are the results of prevailing beliefs, Lambo contends that these beliefs share, in certain instances, a high degree of probability, the best we can expect in any investigation of social-psychological data.

If we examine the process of diagnosis of diseases in these traditional societies, we may be able to rid ourselves of many misinterpretations. The inferences of the traditional healers may not have the cast-iron rigour of a microbiological demonstration, but they register certain important features of the objective situation.[33]

It is sufficient to say that every culture has either intuitive or pragmatic reasons for holding on to a belief. In the case of traditional medicine, there are intuitive and pragmatic grounds for believing in its efficacy.

The aspect of traditional medicine which is explicable and which does not generate so much controversy involves essentially the application of medicinal herbs. Good rainfall and abundant sunshine have enabled the African continent to thrive with a great variety of plants that are good sources of drugs. The African traditional healers have used these medicinal plants effectively so that today traditional medicine remains the widely used for healing in many parts of Africa. This must be so in a continent where the doctor population ratio in 1961 was 25,100 to 1, and 17,500 to 1 in 1972.[34] In Nigeria during 1978 there was one doctor for every 24,000 people and one nurse for every 11,000. In 1970, there was one hospital bed for every 2,150 people in Nigeria, while there was one bed for every 1,350 in 1977. Professor Sofowora had made useful proposals on how to make medicine more accessible and acceptable to people in Nigeria. He advocates the local use of herbs in traditional medicine, their modernization and production in commercial quantities after the efficacy of these herbs had been established in terms of their chemical constituents. In fact, it is suggested that medicinal plants are being exported in large quantities from Africa to Europe and America. His impressive dossier of various medicinal plants of commercial quantity in Africa, including 'Fagara,' a possible treatment of leukemia, suggests a promising future for the development of drugs in Africa.[35]

The plant 'Fagara' has other uses. My late father used it as a tonic. The roots are cut into small pieces and put into a bottle filled with aromatic schnapp. After a few days the liquid turns yellow. From this a small cup, measuring about two to three tablespoons, is taken once a day. It is supposed to rejuvenate the blood. For this reason, it is recommended for adults only. The quality of the blood and the human strength and virility is thought to reduce by age. Under this circumstance, the liquid extracted from 'Fagara' in the form of agbo, as stated above, is seen as giving new strength and virility to the human blood. In certain cases it is called sagbadewe (making an old man become young in blood and strength). It is my belief that the use of 'Fagara' as a tonic is worthy of further investigation, and further research might show that it can serve as an invigorating tonic for the feeble and the aged, and might be useful in increasing the potency of men. The anti-sickening effect of the 'Fagara' has also been noticed in the strength it gives to the teeth when the root is used as a chewing stick. Today, the Drug and Research Unit of the Department of Pharmacognosy, University of Ife, of which Professor Sofowora is the head, is recognized throughout the world. Eminent traditional healers are used as informants, and some of them are invited to give lectures on traditional medicine. Similar units have been established at the University of Lagos and Ahamadu Bello University in Zaria, Nigeria.[36]

It is now becoming increasingly clear that there is need for the integration of traditional with modern medicine. Like its Western counterpart, African traditional medicine requires a long period of training, with traditional healers always making sure that the drugs they develop and use are both effective and safe for humans, after they had been tested on animals. Therefore, it could be said that, like Western medicine, traditional medicine has a definite aim. It is also governed by a methodology and a system of principles which dictate the manner in which the act has to be performed if it is to be effective. One area in which the traditional healers do not make any impact is in surgery, although there are cases of minor surgery covering such operations as circumcision, the partial amputation of fingers, extension of the labia minora, perforation of the ear lobe and of the subseptum in the nose, filing of teeth, subincision of the urethra, and infibulation and excision of the outer genital organs in women. There is surgery for purely therapeutic purposes like the setting of bone fractures, the hemostatic dressing of wounds, the incision of abscesses, and the extraction of guinea worms. Cases of more serious forms of

100

surgery like laparotomy in the treatment of wounds penetrating the abdomen, coupled with suture of the intestine (Abyssinia in Sudan), treatment of traumatic eviscerated conditions by a form of plastic surgery using a calabash disc and suture (also in Sudan), the removal of ovaries, and caesarian section have been reported.[37]

Africans, like Indians and Chinese, have a well preserved system of medicine which is different from the scientific or synthetic medicine of the Western world.[38] Although the efficacy of the Western medicine is not disputed in Africa, India, or China, where it is practiced, the traditional system is highly in demand, especially as millions of patients in both urban and rural areas in these densely populated regions of the world depend on it for treatment. A good example is childbirth. Because of lack of access to modern hospitals, virtually all the babies born in the rural areas in Nigeria and other parts of Africa are delivered at home by herbalists. The herbalists have a system of caring for the baby from the third month in the womb. As soon as the woman is pregnant, the baby is "tied" (secured) in the womb to prevent any miscarriage.[39] At the end of the third month, the woman is given some medication to allow the embryo to receive food properly from its mother. Later, from the fourth month, another medication is given to give strength to the baby as it grows and kicks. From the fifth month the woman continues to take various kinds of agbo aboyun (liquified herbs for pregnant women). The seventh month is particularly important. This is when the baby is given a medication against convulsion (giri) through its mother. It consists of Isu Igbegbe, a mysterious tuber (yam).

In the morning the tuber is found close to the top of the soil. But in the afternoon or evening it moves far down in the ground. The upward and downward movement of this tuber provides an insight into its medicinal value. When obtained, it is cut into pieces, put in a bottle filled with water. A little hole is made on top of the bottle cover and a feather (iko odide) of a beautiful bird known as odide is put through the hole of the bottle cover right inside the bottle, with the sharp end of the feather outside the bottle cover. It is then hung up and must not touch the ground from then on. The liquid from the bottle is taken by the pregnant woman every three days until the whole liquid content is exhausted. The measurement is about half a glass of water each time. A pregnant woman who drinks this potion usually vomits. This is regarded as a normal effect of the medicine. The usual effect of this medicine on the baby when born is the prevention of convulsion, even at high body temperature. After seven months there are medications for easy delivery. One of these is taken regularly by the woman to prevent the placenta from getting too big, and the other, ose dudu (medicinal soap), to make delivery easy. In fact, the effect of these could be seen in women delivering within a few minutes of labor. Many of these deliveries usually take place without the aid of orthodox widwives. Complicated cases are usually taken care of by herbalists who specialize in child delivery.[40]

After birth, treatment of the baby continues. Before and after the advent of Western medicine most Africans were fed on mother's milk and liquified herbs and roots (agbo). The latter, in particular, had some advantages. It was cheap and natural; its potency gave strength and energy to the baby and aided its immunity against unhealthy environments. The problem, however, is lack of knowledge as to the precise quantity a baby should consume at each feeding. By the time the baby is five months old, a specially prepared medicated soap is used to wash its head and face to prevent teething problem. As long as this soap is used about two to three times a week, the baby grows its teeth without problem. One must mention in passing that there are various kinds of fertility drugs, the majority of which are used by females. In certain cases, both husband and wife use the same kind of fertility drug, usually made into stew or soup. But, in general, the men use a popular drug known as aremo, a powdered substance taken with honey. There are various fertility drugs for women.

Something must be said about circumcision of babies, both males and females. In Africa, circumcision is a form of local surgery and is performed by experts when the baby is about eight days to one month old. Usually no anesthesia is used. This may sound strange, but there is no knowledge of anesthesia so that circumcision has to be done early in the baby's life. Even now, the use of anesthesia is not regarded as essential for circumcision. The treatment after circumcision is simple. All that is given as treatment is the liquid extracted from the conical part of a big snail. The liquid which is believed to have an antibiotic effect is dropped on the circumcised part of the male or female organ immediately after circumcision. The circumcised part is subsequently dressed with a fresh leaf. After about seven days, the place heals up and the baby has forgotten everything about the circumcision. He or she is now believed to be properly prepared for his or her future sexual role. Circumcisions done by native doctors on male children are artistically more beautiful than those done by modern doctors in the hospital. But with the development of antibiotics, some educated parents feel more comfortable with modern circumcision just in case some infection should set in which may be difficult to handle in the traditional way. Apart from this unforeseen complication, circumcision in Africa has remained a popular and highly successful aspect of traditional surgery.

The question of infantile mortality in Africa also can be raised. It is common knowledge that infantile mortality in the Western world is lower than in Africa and other undeveloped countries. The reasons for this are obvious: malnutrition, unhealthy environments, disease, and lack of access to medical facilities that are capable of saving human lives during emergencies: surgery, blood transfusions, antibiotics, and oxygen tanks. However, the greatest cause of infantile mortality is poverty which itself is responsible for malnutrition, poor sanitation, lack of clean and clear drinking water, and good housing. These are the preventive aspects of medicare and are generally inadequate in Africa compared to the advanced countries of the world. If these conditions are taken into account, the cause of high infantile mortality in Africa should be well understood. The traditional healers are doing their best, but the only medicine for poor sanitation and malnutrition is a clean environment, clean drinking water, and nutritious food. Such things are not the business of the traditional healers. It is the duty of African governments to take care of these aspects of preventive medicare.

For one reason or the other, the majority of Africans in the rural areas look for treatment of diseases and illness (including mental illness) from the traditional healers who share the same experience with them and so understand their problems. They are thought to be better able than Western doctors to understand the causes of their disease or illness. The thought of getting treatment in the Western orthodox sense usually does not come to their minds. Some even need to be convinced in order to go to a hospital. Because many people die in the hospital, the general feelings among people in the rural areas is that the hospital is a modern road to untimely death. Perhaps faith plays an important role in the acceptance of a particular system of medicine and its efficacy. Every system of medicine has its own limitations. Even with specialized and sophisticated medical systems, people die of diseases and illness. One thing that cannot be denied, however, is that one system of medicine could be better and more efficient than the other in certain specific respects. But both can be integrated for a common good--the general benefit of mankind.

In his special message on behalf of the Organization for African Unity (OAU) to the first international All-African Conference on Health Education held in Lagos, Nigeria on Tuesday, 1 September 1981, the Assistant Executive Secretary of the OAU, A. H. Abdel Razik, said that Africans have faith in traditional healers and herbalists as well as in medicinal plants, and, what is more, that about eighty-five percent of Africans prefer traditional medicine. Professor Sofowora also has said as follows:

> The fact that traditional practitioners provide
> health care for an average of seventy-five percent of
> the population in the developing world (including
> Nigeria) necessitated the consideration given to this
> class of health practitioners. . . . Indeed, WHO,
> UNIDO, UNICEF, OAU, and many other international
> organizations now have big programs for the develop-
> ment of traditional medicine.[41]

A WHO consultation group has described African traditional medicine as follows:

> [It is] the sum total of all knowledge and practices, whether explicable or not, used in diagnosis, prevention, and elimination of physical, mental, or social imbalance and relying exclusively on practical experience and observation handed down from generation to generation, whether verbally or in writing.[42]

This would also be true of India's Ayurveda, an indigenous system of medicine, China's indigenous medicine consisting of Fu Hsi, the practice of Divination, and others concerned with decoction and the use of herbs and their active principles and, of course, acupuncture directed toward the strengthening of the body.[43] All these systems of medicines are rooted in the religious and philosophical beliefs of a people. Thus one scholar writes about Indian traditional medicine: "The Ayurveda, which has the sanctity of the ancient scriptures behind it, has served the sick and suffering humanity on this subcontinent for more than three millennia, and to no small extent continues to serve and save millions of them."[44] The same holds in China, Zaire, North Africa, and some parts of East Africa. In all of these places there is a move toward integrating traditional with modern medicine by the development of research oriented traditional institutions. Beijing's Institute of Indigenous Medicine is an example. So is Delhi's New Center for Ayurvedic Medicine. In these institutions more or less successful efforts have been made to integrate the indigenous systems with modern medicine through a system known as "symbolic traditionalization" whereby new outlooks and methods are inculcated in a traditional guise.[45]

Each new medical miracle, with its sophisticated diagnostic tools, has led to increased costs of medical equipment. The cost of medical equipment has escalated the cost of health care in the advanced countries of the world, particularly the United States, almost beyond tolerance. For poor African countries whose intellectuals and scientists have made little or no contribution to modern technology, sophisticated medical equipment has to be purchased at prohibitive costs. Even if they are purchased they soon break down from the permanent interruptions in the supply of electricity and have to be serviced by experts flown in from Europe or America. Poor Africans! The rising cost of imported drugs in a continent which exports natural herbs to industrialized countries, the prohibitive cost of highly sophisticated medical equipment, and the chronic scarcity of medical doctors all seem to have conspired to force hard choices about who should benefit from modern medical facilities in the face of uneven distribution of wealth and mounting populations. India and China have seen the threat posed by an ever increasing cost of medicare, and have undertaken significant research into traditional medicine with the objective of integrating it with modern medicine. There has been no serious move on the continent of Africa, however, particularly south of the Sahara, to integrate traditional and Western medicines.

In Nigeria, the Nigerian Association of Medical Herbalists (NAMH) and the African Traditional Medical Association (ATMA) made some moves towards mutual recognition, but the Federal government was slow to move. Even the examples of China and India were of no avail. Chief J. O. Lambo, the President of the NAMH, had attended by invitation the OAU Conference in Cairo in 1975 and conferences sponsored by the World Health Organization in 1980. He has also undertaken a tour of the United States where he delivered lectures on African traditional medicine. He received recognition only from one state in Nigeria, his own state of Lagos which in 1979 had established a traditional medical board to regulate the practice of indigenous medicine and promote its growth. Lambo became its chairman. On their own initiatives, NAMH and ATMA award diplomas to candidates who have successfully undergone training and passed certain examinations in traditional medicine. After a prescribed examination, NAMH awards a fellowship certificate to distinguished herbalists of twenty years experience who then can add the prefix FNAMH (Fellow of Nigerian Association of Medical Herbalist) after their names. The FNAMH is the highest degree and is seen as equivalent to the academic Ph.D. degree. Thus Fellows of Nigerian Association of Medical Herbalists can call themselves native doctors. A certificate of post-graduate fellowship issued by the NAMH reads as follows:

> This is to certify that Chief, Mr., or Alhaji
> _____, has fully undergone a postgraduate course in
> the faculty of herbalism as prescribed by the
> association's examining board in accordance with the
> articles of association 6-12 of the memorandum and
> articles of NAMH and has successfully passed the
> final examination in the first, second, or third
> class and is therefore awarded this certificate of
> fellowship as laid down in the Constitution of the
> association, page 3, clause 9.[46]

These efforts give credit to the initiative of the native doctors themselves. However, the tragedy in the Nigerian situation is that the Western-trained Nigerian doctors make integration difficult if not impossible. The traditional healers see such actions as motivated either by jealousy or hypocrisy.[47] It is probably both but more of the latter, because many of the Western-trained doctors themselves privately patronize the traditional healers for protective or preventive medicine as well as the treatment of some illness including those which modern medicine is unable to cure.[48]

We are now in a position to compare Euro-African philosophers with Euro-African medical doctors who have abandoned their traditional values, beliefs, and cultures for Western values, beliefs, and cultures almost without remainder. Very often, Euro-African doctors either fail, or refuse, to see what some of their Western counterparts have to say about traditional medicine. A Westerner, Benjamin Walker, said the following:

> Although most orthodox physicians have tended to scorn the methods used by primitive healers as so much superstition, there is today a growing appreciation of the genuine value of some of these methods. Centuries of experience have gone into the evolution of formative healing, and their cures, both psychological and medicinal, have often been very remarkable.[49]

Yet another commentator has said that African traditional medicine has made a noteworthy contribution to medical knowledge in Europe.[50] Like Euro-African philosophers, however, the Euro-african medical doctors appear to be ignorant of the evolution of cultures, ideas, philosophy, and scientific and medical knowledge. They fail to recognize that even the most sophisticated philosophical, medical, and scientific theories today started as myths.[51] Instead of promoting research that might lead to further growth of knowledge in the field of traditional medicine, many African doctors, especially Euro-African doctors, ridicule any effort made towards that direction. But then, as Lambo said:

> The conflict is accentuated by a severe degree of ambivalence to traditional medicine by these practitioners of Western medicine whose cultural background, in spite of their training, is deeply rooted in traditional magico-religious or spiritual explanation of illness.[52]

One also does not know whether theirs is a case of inferiority complex, or a lack of the spirit for national development which their Indian and Chinese counterparts admirably possess. In their fear of offending the analytic spirit of Western philosophical tradition, the Euro-African philosophers saw nothing philosophical about their own philosophy. For much the same reason the Euro-African doctors saw nothing of note in a system of traditional medicine by which many of them were safely delivered, nurtured, and successfully treated until they grew up to be medical doctors.

It is important to point out here that during the question and answer period that followed the delivery of my Fulbright Hays lecture, a member of the audience, an American wife of a Nigerian in the United States, confirmed the efficacy of African traditional medicine from her own experience. She reported that she

106

did not believe in African traditional medicine until she was taken to a native doctor during one of her visits to Nigeria. She had a long history of migraine headaches which had resisted cure in the United States. Her experience showed that she was cured within minutes by a Nigerian native doctor among the Yoruba, at little or no cost. Surely, in the face of the escalating cost of medicare the integration of traditional with modern medicine could force down the cost of treatment of various diseases and illness that do not require the use of highly sophisticated equipment. People would prefer an effective cure for migraine headaches gotten at once by the use of traditional medicine and at a very low cost than a cure by modern medicine at a prohibitive cost. If one could get a treatment from acupuncture for fifty dollars, and a similar treatment from modern surgery for five hundred dollars, the majority of people would prefer the former. What should be regarded as important is not the particular system but its efficacy. Insofar as both systems lead to a cure, the lower the cost the better it is. The crucial thing is complete cure (iwosan).

I see no reason therefore for African medical doctors to look down on traditional healers or physicians and treat the African system of traditional medicine with cries of mockery. Certainly African traditional medicine is capable of improvement. The possibility of such improvement lies with the Western-trained medical doctors. They should show more interest in research into traditional medicine and see to it that African traditional physicians acquire a basic training in modern scientific methods. This could lead to the establishment of modern institutes of traditional or native medicine similar to those in Beijing and New Delhi where all effort is being made to integrate the traditional with modern systems of medicine. If this is successfully done, African traditional medicine which had been derided by Western trained medical practitioners may, in the face of integration and modernization, sometime in the future become the seeds that are likely to proliferate into the fruits of Western medical science. There would then be a need to develop a system of bio-ethics which would attempt to clarify ethical issues for both the orthodox and traditional physicians as well as the public. In this way, African traditional medicine would take its place in the world of medical science and traditional African physicians gain international respectability.

Chapter 6

CONCLUDING REMARKS

It is the bitter truth that Africans, particularly members of governments, intellectuals, and policy makers, want things to be done for them. They want philosophy, medicine, science and technology, consultancy services, development planning, and strategic policies to be done by foreigners. The Chinese and Indians developed and are practicing traditional medicine without waiting for an official stamp of approval from Western cultures. They are also doing their philosophy which many Westerners read and study with sustained interest. But Africans seem to be waiting for others to help them do research and write on African traditional medicine. They also seem to think that African philosophy must of necessity conform to the Western analytic concept before it is acceptable as philosophy.[1] I am not aware that Chinese or Indian philosophers regard their philosophies in the same way as Africans think of African philosophy.

If the current trend of intellectual abuse of certain philosophically and medically significant aspects of African traditional cultures and thought systems continues unabated, one would not be surprised to see the majority of the textbooks on African philosophy and on traditional medicine written by Europeans and Americans for African professors of philosophy and medical doctors. The most valuable textbook for a course on African philosophy given during my Fulbright research program in the United States in 1984 was one edited by Richard A. Wright, an American professor of philosophy at the University of Toledo, Ohio. To my knowledge, Wright's book was the first to be titled African Philosophy. This was at a time when many Africans were busy debating whether African philosophy existed.[2] This is no surprise: many African nations hear the news of their country first from the British Broadcasting Corporation (BBC) or the Voice of America!

Africans should take pains to examine their traditional beliefs, critically discuss those that are inconsistent with other beliefs, and use their own philosophical training to judge which of these inconsistent beliefs must be rejected. In so doing they must not reject an African belief merely because it is inconsistent with Western belief.

Materials already exist for the development of African philosophy and traditional medicine. What is needed is the super-imposition of critical discussion on these materials, however crude some of them may be. In the absence of critical discussion the mind could be led to one of the two extremes-- wholesale rejection or wholesale acceptance of our particular belief system. In accepting or rejecting an idea, it would be wrong to look at it mainly from the point of view of modern progress and development or in terms of science and technology.

What I have tried to do in this book is suggest that it will never be the case that traditional thought is incapable of contributing to knowledge of any kind, even scientific knowledge. African traditional thought is neither dogmatic nor opposed to change. Nonetheless, many aspects of African traditional thought and medicine are incapable of contributing to modern scientific knowledge. What I advocate, therefore, is not an uncritical acceptance of African traditional beliefs, but a critical acceptance of some, and a critical rejection of others.

I do believe that Africans, from their native intelligence, can make their own contributions to philosophy, science, and medicine. At present, many African nations, particularly Nigeria, talk about transfer of technology as if the whole of Western technology were to be moved from Europe and the United States to Africa. Technology, as I have argued elsewhere, cannot be transferred.[3] Japan and China borrowed from Western technology to improve on their own. Even if the so-called transfer of technology were possible, Africa would carry with it the transfer of the good as well as evil aspects of Western technological cultures and the purely materialistic conception of the person.

Finally, we must anticipate that the ever-rising cost of health care in the advanced countries may force people to look for alternative systems of health care delivery. This is a situation that would inevitably lead to the integration of expensive orthodox medicine with cheap but effective traditional medicine. Chinese acupuncture is already making an impact on the world of medicine. The secrecy and mystery surrounding African traditional medicine can only be unraveled by modern research and useful publications. In this way, Africans can make their own contributions to medical knowledge. The final integration of traditional African medicine with modern medicine would surely provide an alternative to some aspects of Western medicine.

My hope is that this book has effectively demonstrated the potentials for the development of African philosophy as well as traditional medicine; that it is the duty of contemporary African philosophers and medical doctors to show more interest in these fields and develop them through research; that if they fail to do this, others would surely do it for them.[4]

APPENDIX

AFRICAN TRADITIONAL PRACTICE AMONG

THE YORUBA OF NIGERIA

Table 1

ORGANIC ILLNESS/DISEASE

Kinds of Traditional Healers/Native Doctors

---

1. General: Herbalists (onisegun, known as the servants of orunmila)

2. Specialists: Based on the same principles and practice, but with more specialization in diagnostic methods, treatment and medication of particular areas of traditional medicine like obstetrics and gynecology (a toju awon aboyun, igbebi omo ati iwosan agan tabi arun obirin), bone setting (eguntito), pediatricians (a toju awon omo owo), etc.

For the psychiatrist (awowere/a toju were/asinwin).
See Table 3: Functional illness/disease, below.

Diagnostic Methods

Native (clinical) testing procedures
Semeiology/symptomatology
Occasional divination

Treatment Purposes

Curative
Restorative
Preventive
Protective

Table 1
(continued)

Methods of Treatment, and Medications

Medicinal Herbs & Incantations

(a) Natural/scientific

    (i) Internal applications: Drugs - Agbo: liquid from natural herbs, roots, leaves, bark of trees; Agunmu: ground or powdered natural herbs, roots, leaves, bark of trees and animal substances; Igbere (powdered, native injection): ground or powdered natural substances from leaves, herbs, roots, bark of trees, animal or other natural substances

    (ii) External applications: Medicinal ointments from natural substances (ebe) and lotions (ipara eg Adin); native medicinal soap (ose dudu)

(b) Magical

    (i) Power or authority of words (like exousia or Dunamis), with magical horns as its symbols. Used with incantations.

    (ii) Ofo (ordinary incantations) may include curse or imprecation (epe) on suspected cause of illness.

    (iii) Sacrifices or propitiations, as may be directed by Ifa.

    (iv) Amulets (isora) like magical rings (iroka), magical belts (onde) and other kinds of objects inscribed with magic incantations or symbols.

114

Table 2

ORGANIC/FUNCTIONAL ILLNESS/DISEASE

Kinds of Traditional Healers/Native Doctors

Babalawo (Diviners-Ifa priests)

Diagnostic Methods

Ifa Divination

Treatment Purposes

Curative
Preventive
Protective

Methods of Treatment, and Medications

Oral Medicine

(a) Magical:

    (i) Odu Ifa: chanting of verses
    (ii) Incantations

(b) Sacrifices/propitiations (etutu, ebo sometimes believed to be more powerful than ogun (herbal medicine). Often used to aid the process of complete cure (iwosan), i.e. treatment from the root cause of illness or disease.

(c) Occasional use of herbal medicine, e.g., ewe ifa (Ifa leaves), lyerosun, or lye (a yellowish powder extracted from the tree, Irosun. Used for Ayajo (protective/ preventive/ curative), spread out on Ifa or any tray, marked with the sign of a particular odu Ifa, collected together, eaten together with obi (Kola nut) or may be mixed with other things e.g. epo pupa (palm oil) and may be used when mixed with oil, as ipara (medicinal ointments).

115

Table 3

FUNCTIONAL ILLNESS/DISEASE

Kinds of Traditional Healers/Native Doctors

Awowere (Psychiatrist): Were or iwin is an extreme form of mental disorder. But were/iwin is also in degrees. Those who parade the streets naked or half naked with hair unkempt and have slept in the market places are usually not accepted for treatment by traditional healers because they think they are incurable, especially if they have slept in the market places. These may be called the most extreme form of were (were paraku) who are better kept in the lunatic asylums.

Diagnostic Methods

(a) Yoruba Aetiology

(b) Ifa Divination: Owonrinwere, iworiwo osa kan, owonyewere. Reveal the following:

1. Physical cause: e.g. ode ori (a disease in the brain caused by certain germs), poisoning, misapplication of drugs or igbere.
2. Wilful Maleficia (Aiyekunrin and Aiyebirin, or simply Aiye/Alaiye)
   a. Occult: epe (curse or imprecation), ase, ayajo
   b. Spiritual cause: Aje (witches) usually females (aiyebirin), with their private parts as the sources of their powers.
3. Metaphysical cause: associated with family gods, e.g. ogun (god of iron), osun (god of water), sango (god of thunder), etc.
4. Hereditary factor: Mental disorder as transmitted from one generation to another.
5. Reward for evil deeds.
6. Metaphysical inner head (ori inu) or guardian spirits--the bearer of human destiny.

116

Table 3
(continued)

Others

7. Esoteric: The totality of what is known as a person is believed to consist of elements that are connected with a chain. A rupture in the chain of the elements, and depending on the point of rupture, may lead to insanity. This is a complex doctrine intelligible only to the initiated.
8. Hemp smoking.

Treatment Purposes

Curative (sometimes permanent in nature)
Restorative

Methods of Treatment, and Medications

Usually at the psychiatrist's home, under dangerous condition which the healer controls with ofo (incantation) or ase to bring patient under control so as to discuss, cooperate and accept medications readily.

(a) Magical

    (i) Ofo/asisan/ayajo/afose (incantations) directly targeted against the patient's brain.
    (ii) Sacrifices/propitiations

(b) Herbal

    (i) Igbere (powdered native injection applied to the head or other places.
    (ii) Some powerful sleeping potion: liquid extracted from the roots, herbs, bark of trees or leaves, e.g., roots of a tree known as Asofeyeje.
   (iii) Agbo: liquid extracted from plant substances for drinking or bathing.
    (iv) Agunmu
    (v) Medical soap (ose dudu)

Table 3
(continued)

(b) <u>Herbal</u> (continued)

    (vi) Medical ointment (<u>ipara</u>)
    (vii) The use of medicinal canes has been discouraged and so
          very rarely is used. The cane was to drive away evil
          spirits. <u>Igbere</u> and incantations are used instead.[*]

_____

[*] For a detailed discussion on the treatment of mental disorder in the Yoruba Society, see M. Akin Makinde, "Cultural and Philosophical Dimensions of Neuro-Medical Sciences," <u>Nigerian Journal of Psychiatry</u>, September 1987, pp. 85-100.

Chapter 1

1. See Maurice De Wulf, Scholastic Philosophy: Medieval and Modern, translation by Peter Coffey (New York: Dover, 1956), Ch. 1. See also Arthur Hyman and James J. Walsh, Philosophy in the Middle Ages (New York: Harper & Row, 1967).

2. St. Augustine (354-430 A.D.) was born in Thagaste in North Africa. As a Christian philosopher he considered knowledge of any kind as a function of the soul. See R. A. Markus, "Augustine, St." in Paul Edwards, ed., The Encyclopedia of Philosophy (New York: Macmillan, 1972) 2 vols., pp. 198-206; H. I. Gilson, St. Augustine and His Influence Through the Ages (London: 1958); E. Gilson, The Christian Philosophy of St. Augustine, translation by L. E. M. Lynch (London: 1961); R. A. Markus' "St. Augustine" in D. J. O'Conner, ed., A Critical History of Western Philosophy (New York: The Free Press, 1964), pp. 80-97.

3. According to Karl Popper, "Western Philosophy consists very largely of world pictures which are variations of the theme of body-mind dualism, and of problems of method connected with them." See his Objective Knowledge: An Evolutionary Approach (Oxford: Clarendon Press, 1972), p. 153.

4. See Ernest Nagel, "In Defense of Atheism," in Paul Edwards and Arthur Pap, eds., A Modern Introduction to Philosophy (New York: The Free Press, 1965), pp. 460-72.

5. See Henry Olela, "The African Foundations of Greek Philosophy," in Richard A. Wright, ed., African Philosophy (New York: University of America Press, 1983), third edition, pp. 77-92, and Lucinay Keita, "The African Philosophical Tradition," in the same volume, pp. 57-76.

6. The philosophical works of Francis Bacon, Thomas Hobbes, and John Locke did much to bring about the political, philosophical, cultural, and scientific changes that started later in the 18th century (the Age of Reason). Francis Bacon's writings were particularly influential in the development of British Empiricism. His works also had some influence on Sir Isaac Newton and the Victorian scientists and methodologists. See Francis Bacon, De Interpretatione Naturae, Essays (1597), Advancement of Learning (1605), Novum Organum (1650), and the New Atlantis

(1624); Thomas Hobbes, The Elements of Law: Natural and Political published in two parts as Human Nature and De Corpore Politico (1650), Leviathan (1651) and Body, Mind and Citizen; John Locke Essay Concerning Human Understanding (1689) and Two Treatises of Government (1690).
7. George Basalla, William Coleman, and Robert H. Kargon, eds., Victorian Science: A self-portrait from the Presidential Address of the British Association for the Advancement of Science (New York: Doubleday, 1970). Quotation from the authors' introduction to W. C. Carpenter's on "Man and the Interpreter of Nature," p. 13.
8. Karl Marx, The Wisdom of Karl Marx (New York: Philosophical Library, 1967). Forward by Stockhammer, section B. This small book has no page numbers. Names and ideas are numbered in alphabetical order. For the purpose of easy reference we refer to each alphabet as a section.
9. Basalla, Victorian Science, p. 17.
10. John Stuart Mill, A System of Logic: Ratiocinative and Inductive (London: Longman, 1970), 8th edition, especially Book VI on "The Logic of the Moral Science." Mill's work as indicated above has led to what we now know as "The Methodology of the Social Sciences," an attempt to subject moral, social, economic, and political study to a quasi-scientific investigation.
11. Basalla, Victorian Science, p. 19.
12. M. Akin Makinde, "The World and its Enemies: A Philosophical Perspective," presented at "The International Conference on George Orwell's Nineteen Eighty-Four and its Implication for life Today, The Ohio State University, Columbus, Ohio, U.S.A., May 4-6, 1984. See also M. Akin Makinde, "Technology Transfer: An African Dilemma" in John W. Murphy, Algis Mickunas and Joseph J. Pilotta, eds., The Underside of High Technology, New York/London: Greenwood Press, 1986, pp. 177-189. A revised version of paper delivered for the College of Engineering, Ohio University, Athens, Ohio, 15 May 1984.
13. John S. Mbiti, African Religions and Philosophy (New York: Grove Press, 1961).
14. Kwame Gyekye, "African Religions and Philosophy," in Second Order: An African Journal of Philosophy, Vol. IV, No. 1, 1975, pp. 86-94, and John A. A. Ayoade, "Time in Yoruba Thought," in Wright, African Philosophy, pp. 93-112.
15. J. O. Sodipo, "Notes on the Concept of Cause and Chance in Yoruba Traditional Thought," Second Order, Vol. II, No. 2, 1973, pp. 12-20, and Helaine K. Minkus, "Causal Theory in Akwapim Akan Philosophy," in Wright, African Philosophy, pp. 113-48.
16. E. G. Parrinder, African Traditional Religions (London: Hutchinson's University Library, 1962).
17. Tai Solarin, social critic and critic of christian theology. His public atheistic pronouncements led to a

conflict between him and the Roman Catholic Bishop of Lagos, Nigeria, Bishop Christopher Okogie.

18. M. Akin Makinde, "Immortality of the Soul and the Yoruba Theory of Seven Heavens (Orun Meje)," Journal of Cultures and Ideas, Vol. 1, December 1983, pp. 31-59.
19. See the following from Wright, African Philosophy: Kwame Gyekye, "Akan Concept of a Person," pp. 199-212; Ifeanyi A. Menkiti, "Person and Community in African Traditional Thought," pp. 171-82; Benjamin Eruku Oguah, "African and Western Philosophy: A Comparative Study," pp. 212-26; Richard C. Onwuanibe, "The Human Person and Immortality in IBO (African) Metaphysics," pp. 182-98. See also M. Akin Makinde, "An African Concept of Human Personality: The Yoruba Example," Ultimate Reality and Meaning, Vol. 7, No. 3, 1984, pp. 189-200.
20. See Onwuanibe in Wright, African Philosophy; Wande Abimbola, Ifa: An Exposition of Ifa Literary Corpus (Ibadan: Oxford University Press, 1976) and his La Notion de Personne en Afrique Noire (Paris: Centre National de la Recherche Scientifique, 1971).
21. See Wande Abimbola, Ijinle Ohun Enu Ifa, "Apa Kini," or Vol. 1 (Glasgow: Collins, 1968); Ijinle Ohun Enu Ifa, "Apa Keji," or Vol. 2 (Glasgow: Collins, 1969); Sixteen Great Poems of Ifa (Niamey: UNESCO, 1975); Ifa Divination Poetry (New York: Nok Publications, 1977), Awon Oju Odu Mereerindinlogun (Ibadan: Oxford University Press, 1977).
22. M. Akin Makinde: "A Philosophical Analysis of the Yoruba Concepts of Ori and Human Destiny," International Studies in Philosophy, Vol. XVII, No. 1, 1985, pp. 53-69.
23. M. Akin Makinde, "Ifa as a Repository of Knowledge," a paper presented at the XVIIth World Congress of Philosophy, Montreal, August 21-27, 1983. Also in ODU: A Journal of West African Studies, No. 23, 1983, pp. 116-121.
24. M. Akin Makinde, "Formal Logic and the Paradox of Excluded Middle," International Logic Review, No. 15, June 1977, pp. 40-52.
25. Plato, The Republic, translation by Paul Shorey in Plato's Collected Dialogues, edited by Edith Hamilton and H. Cairns, New York: Bollingen Foundation, 1964.
26. Pierre Simon Laplace, Introduction to his Theory of Probability, a translation from the first French edition (1812) under the title Theorie Analytique des Probabilities. Laplace's Omniscient Intelligence is often quoted by philosophers of science. A popular exposition is contained in Laplace's later work, Essai Philosophique (1814) which has been published in English under the title A Philosophical Essay on Probabilities (New York: Dover Publications, 1952).
27. H. M. Patcher, Paracelsus: Magic into Science (New York: Collier Books, 1961), pp. 78ff. From Paracelsus' discussion it appears that the scientist is a "magus" and

vice-versa. Bombastus Paracelsus was a famous astrologer, alchemist and miracle worker. The notion of "Sagacious Philosophy" has since been widely used, though wrongly I believe, to describe some aspects of African Philosophy. See H. Odera Oruka, "Sagacity in African Philosophy," International Philosophical Quarterly, December 1983, and P. O. Bodunrin, "The Question of African Philosophy," Philosophy, Vol. 56, 1981, pp. 161-79.

28. William Stanley Jevons, The Principles of Science: A Treatise on Logic and Scientific Method (New York: Dover, 1958), p. 197. A nineteenth century British methodologist, Jevon's book was first published in 1873.
29. See note 2 above.
30. Paulin Hountondji, African Philosophy: Myth and Reality (Bloomington: Indiana University Press, 1983). See also his interview in the magazine West Africa, August 22, 1983, p. 1955. The term "ethnophilosophy" was first used by the late Kwame Nkrumah in a thesis he wrote as a student in the United States of America. See P. O. Bodunrin, "The Question of African Philosophy," above note 27, p. 161.
31. Hountondji, "African Philosophy" in West Africa, above note 30, p. 1955.
32. Justice Adewale Thompson, African Beliefs: Science or Superstition? (Ibadan: Newton House Publications, 1977), p. 264.
33. The idea of ethno-medicine is implicit in Pierre Huard's article "Western Medicine and Afro-American Ethnic Medicine" in F. N. L. Poynter, ed., Medicine and Culture (London: Wellcome Institute of the History of Medicine, 1969), New Series, Vol. XV, pp. 211-37.

Chapter 2

1. Douglas Hubble, "Medicine and Culture" in F. N. L. Poynter, ed., Medicine and Culture (London: Wellcome Institute of the History of Medicine, 1969), p. 79.
2. Hubble, "Medicine and Culture," p. 79.
3. Geoffrey Vickers, "Medicine's Contribution to Culture," in Poynter, Medicine and Culture, p. 5.
4. Vickers, "Medicine's Contribution," p. 5.
5. Popper, Objective Knowledge, Chs. 3, 4.
6. For Popper on the interesting distinction between Einstein and Amoeba, see Popper, Objective Knowledge, pp. 24-25 and p. 247.
7. See Langdon Gilkey, "The Religious Dilemmas of a Scientific Culture: The Interface of Technology, History, and Religions," in Donald M. Borchert and David Stewart, eds., Being Human in a Technological Age (Athens: Ohio University Press, 1979). See also John G. Burke, ed., The New Technology and Human Values (Belmont, CA: Wadsworth, 1972);

Bertrand Russell, The Scientific Outlook (New York: W. W. Norton, 1962), and Jonathan Schell, The Fate of the Earth (New York: Avon Books, 1982).

8. Bertrand Russell, History of Western Philosophy (London: George Allen & Unwin, 1962), 18th edition, p. 14.
9. Russell, History, p. 7.
10. See David Stewart and H. Gene Blocker, Fundamentals of Philosophy (New York: Macmillan, 1982), pp. 36-41.
11. Peter Caws, Philosophy of Science (Princeton: Van Nostrand, 1966), p. 32.
12. B. L. Whorf, Language, Thought and Reality (New York, 1956), cited in Caws, Philosophy.
13. The conflict between British empiricism and Continental rationalism with respect to Cartesian intuition and deduction and Bacon's empiricism and induction is well known. See A. E. Taylor, Francis Bacon (London: Oxford University Press, 1926) and M. Akin Makinde, "John Stuart Mill's Theory of Logic and Scientific Method as a Rejection of Hume's Scepticism with Regard to the Validity of Inductive Reasoning." Unpublished dissertation, University of Toronto, 1974, pp. 149-51, Part II, Ch. 3. Sections. IV-V, and Ch. 4, Section 1.
14. See Richard von Mises, Positivism (New York: Dover Publications, 1968), p. 51. The expression "critical grammar" is also used instead of "logical grammar."
15. The colloquium in which Professor Wole Soyinka expressed this view was the most intellectually significant of the 1977 Festival of Arts & Culture, the remaining contributions being mainly in the areas of music and dance. For a good discussion on this issue, see M. Akin Makinde, "The Possibility of an African Continental Language: A Philosophical Analysis," Journal of African Studies (UCLA), forthcoming.
16. The two European languages in Africa are spoken in the former English and French colonies in West, Central, and East Africa. These languages have become the official languages of academic instruction, trade, and all forms of business as well as the official languages of parliaments. Contemporary African philosophers are divided along this linguistic line. Thus the philosophers from former Anglophone British colonies lean towards British analytic philosophy.
17. See Robin Horton, "Traditional Thought and the Emerging African Philosophy Department: A Comment on the Current Debate," Second Order, Vol. VI, No. 1, 1977, pp. 64-80. See also P. O. Bodunrin, "The Question of African Philosophy," Philosophy, Vol. 56, 1981, pp. 161-79, reprinted in Wright, African Philosophy, pp. 1-24. For a rejoinder to Robin Horton's paper above see Barry Hallen, "Robin Horton on Critical Philosophy and Traditional Thought," Second Order, Vol. VI, No. 1, 1977, pp. 81-92.

Chapter 3

1. William P. Alston and Richard B. Brandt, The Problems of Philosophy (Boston: Allyn & Bacon, 1967), Introduction. See also James K. Feibleman, Understanding Philosophy (New York: Horizon Press 1973).
2. Russell, History, p. 13.
3. Russell, History, p. 13.
4. Russell, History. See also Bertrand Russell The Art of Philosophizing (New York: Philosophical Library, 1968), pp. 33-35. He recognizes that philosophy, as conceived by the Greeks, was as a way of life. "A complete philosophy," says Russell, "will have a conception of the ends to which life should be devoted, and will in this sense be religious," although he also recognizes the scientific aspect which is the pursuit of knowledge.
5. Russell, The Art, pp. 33-35.
6. Russell, History, pp. 13-14.
7. Russell, History, p. 13.
8. Russell, History, pp. 13-14.
9. C. M. Joad, Philosophy (London: Richard Clay, 1974), p. 39.
10. Berkeley was a philosopher who tries to make sense out of nonsense. However, Berkeley's subjective idealism has been criticized and rejected by many philosophers, the most prominent of which is G. E. Moore. Logical positivists take delight in making nonsense out of sense. See Rudolf Carnap, "Elimination of Metaphysics through Logical Analysis of Language," in A. J. Ayer, ed., Logical Positivism (New York: The Free Press, 1959). For the manifesto of logical positivism, see A. J. Ayer, Language, Truth and Logic (London: Victor Gollanz, 1970), 18th ed. For the critics of logical positivism see A. C. Ewing, "Meaninglessness," in Mind, 1937, and W. T. Stace, "Metaphysics and Meaning," in Mind, 1935. Both articles are reprinted in Paul Edwards and Arthur Pap, eds., A Modern Introduction to Philosophy (New York: The Free Press, 1965), pp. 705-14 and 694-704 respectively.
11. Friedrich Waismann, "How I See Philosophy," in Ayer, Logical Positivism, p. 375.
12. John Gottlieb Fichte, quoted in Theodor Oizerman, Problems of the History of Philosophy (Moscow: Progress Publishers, 1973), translated by Robert Daglish, p. 225.
13. George Chatalian, "Philosophy, The World and Man: A Global Conception," Inaugural Lecture delivered at the University of Ife, Ile-Ife, Nigeria, on 28 June 1983, Part IV, Sec. 2.
14. R. G. Collingwood quoted in Lionell Rubinoff, Collingwood and the Reform of Metaphysics (Toronto: University of Toronto Press, 1970), p. 150.
15. Rubinoff, Collingwood, p. 124. See also Collingwood's Speculum Mentis (Oxford: Clarendon Press, 1924), p. 260.

16. Rubinoff, Collingwood, p. 151. See also Collingwood, Speculum Mentis, p. 309.
17. Waismann, "How I see Philosophy," p. 375.
18. Russell, The Art, p. 10.
19. Russell, The Art, p. 10. Awolowo's philosophical thought is similar to Plato's in this respect.
20. Oizerman, Problems, in note 12 above, p. 225.
21. Oizerman, Problems, pp. 228-29.
22. Bertrand Russell, History, p. 14.
23. See David Hume, A Treatise of Human Nature (Oxford: Clarendon Press, 1967, L. Selby-Bigge edition and his Enquiry Concerning the Human Understanding, Selby-Bigge edition (Oxford: Clarendon Press, 1962).
24. Lucinay Keita, "The African Philosophical Tradition," in Wright, African Philosophy, pp. 57-76.
25. At least more than a third of African Universities have established departments of Philosophy, either combined with Religious Studies or as separate departments. The following divisions are notable: North Africa, West Africa, Central Africa, East Africa, and South Africa (Zimbabwe, Lesotho, etc.), with each country in the divisions having at least from one to three universities. In Nigeria alone there are at present about twenty-four universities (both Federal and State) in all the nineteen states. Nine of these universities teach philosophy either in separate departments or in combined departments of religious studies and philosophy. The Federal Universities of Ife, Ibadan, Lagos, and the State Universities at Ekpoma in Bendel State, the Obafemi Awolowo University (OAU) at Ado Ekiti in Ondo State, and Ogun State University at Ago-Iwoye have separate departments of philosophy while the Federal Universities of Port Harcourt in Rivers State, Calabar in Cross-Rivers State, Benin in Bendel State, and the University of Nigeria at Nsukka in Anambra State combine philosophy with religion as a department. The Nigerian Philosophical Association meets every two years for presentation of philosophical papers from within and outside the continent. There are at present two notable Journals of Philosophy: Second Order published at the University of Ife, and the Nigerian Journal of Philosophy published at the University of Lagos.
26. Four people usually are known as "the paradigmatic individuals": Socrates, Buddha, Confucius, and Jesus Christ. Only Confucius wrote anything. The thoughts and teachings of the other three were written by their disciples. See Karl Jaspers, "Socrates, Buddha, Confucius, Jesus: The Paradigmatic Individuals," in Hannah Arendt, ed., The Great Philosophers (New York: Harcourt, Brace & World, 1962), Vol. 1, translated by Ralph Manheim.
27. Russell, History, p. 14.

28. The origin of this term has been traced to the late Kwame Nkrumah of Ghana. Its current wide usage among contemporary African philosophers is probably due to the Francophone philosopher, Paulin Hountondji.

29. Although there are no records from the time of the visit of Pythagoras to Egypt, a record of his probable date of birth exists. Since St. Augustine lived between 354-430 A.D., it seems reasonable to suggest that some philosophy was done and written between 570 B.C. and 430 A.D. in North Africa.

30. See R. A. Markus, "Augustine, St.," in Paul Edwards, ed., The Encyclopedia of Philosophy (New York: Macmillan, 1972), Vol. 1, p. 200.

31. See M. Akin Makinde, "Immortality of the Soul and the Yoruba Theory of Seven Heavens," in Journal of Cultures and Ideas, Vol. 1, No. 1, December 1983, pp. 31-59.

32. Placide Tempels, Bantu Philosophy (Paris: Presence Africaine, 1959), pp. 167-69.

33. See W. A. Hart, "The Philosopher's Interest in African Thought: A Synopsis," in Second Order, Vol. I, No. 1, 1972.

34. Russell, History.

35. Tempels, Bantu Philosophy, pp. 168-69.

36. For an article on Kalabari thought, see Robin Horton: "The Kalabari World-View: An Outline and Interpretation," in Africa, Vol. XXXII, No. 3 1962, pp. 197-220. Kalabari is a small linguistic community in Southeastern Nigeria.

37. Kwasi Wiredu used the word "orientation" in his paper entitled "On an African Orientation in Philosophy," Second Order, Vol. I, No. 1, 1972, pp. 3-13.

38. For some of this debate, see Wiredu "On an African Orientation"; Kwasi Wiredu, "How Not to Compare African Traditional Thought with Western Thought," in Ch'Indaba, No. 2, July/December 1976, reprinted in Wright, African Philosophy; H. Odera Oruka, "Mythologies as African Philosophy," East African Journal, Vol. 9, October 1972; H. Odera Oruka, "The Fundamental Principles in the Question of African Philosophy," Second Order, Vol. IV, 1975; E. A. Ruch, "Is There an African Philosophy?" Second Order, Vol. III, No. 2, 1974; A. Finazz; "Una Filosofia Africana?" Africa (Rome), Vol. 29, 1974; Henri Maurier, "Do We Have an African Philosophy?" in Wright, African Philosophy; Richard A. Wright, "Investigating African Philosophy," in Wright African Philosophy; P. O. Bodunrin, "The Question of African Philosophy," in Wright, African Philosophy, originally in Philosophy, Vol. 56, 1981; Robin Horton, "Traditional Thought and the Emerging African Philosophy Department: A Comment on the Current Debate," Second Order, Vol. VI, No. 1, 1977; Paulin J. Hountondji, "African Philosophy: Myth and Reality?" Thought and Practice, Vol. 1, 1974; Paulin J. Hountondji African Philosophy: Myth and Reality (Bloomington: Indiana University Press, 1983).

39. See W. A. Hart, "The Philosopher's Interest in African Thought," Second Order, Vol. I, No. 1, 1972, pp. 43-52.
40. See M. Akin Makinde, "Robin Horton's 'Philosophy': An Outline of Intellectual Error," an unpublished paper (105 pp.) presented as a philosophy colloquium, Faculty of Arts, University of Ife, on 20 June 1978, and our notes 18, 19, 22 and 23 (in chapter 2) and 53, 55, 58, and 66 in the present chapter.
41. Makinde, "Robin Horton's Philosophy."
42. John Locke made a similar point against Aristotle and his so-called invention of logic. In his argument Locke tried to show that men reason correctly without the development or aid of logic. He therefore wondered: "If syllogism must be taken for the only proper instrument of reason and means of knowledge, it would follow that before Aristotle there was not one man that did or could know anything by reason. . . . But God has not been so sparing to men as to make them barely two-legged creatures, and left it to Aristotle to make them rational." John Locke, An Essay Concerning Human Understanding (Oxford: Clarendon Press, 1969), Bk. 4, pp. 346-47.
43. Popper, Objective Knowledge, pp. 120-22.
44. See Popper, Objective Knowledge, p. 126. See also Popper's Conjectures and Refutations: The Growth of Scientific Knowledge, New York: Harper & Row, 1963, Ch. 15, and The Open Society and Its Enemies, London: Routledge and Kegan Paul, 1952 in addendum to Vol. II: "Facts, Standards and Truth: A Further Criticism of Relativism."
45. See Lucian Levy-Bruhl, Primitive Mentality (Boston: Beacon Press, 1966) (reprint of 1923 edition), Lucian Levy-Bruhl, Now Natives Think (New York: Alfred Knopf, 1925), and Horton, "Traditional Thought."
46. See Nader Chokr, "Mankala: Wari & Solo: A Study of a Board Game Played by African People," the Archives of Smithsonian Institution, Washington, D.C., 1985. See also C. Zaslavski, African Counts: Number and Pattern in African Culture (Boston: Prindle, Weber & Schmidt, 1973); M. B. Nsimbi, Omweso, A Game People Play in Uganda (Los Angeles: University of California, 1968), African Studies Center, Occasional Paper; E. M. Avedon and B. Sutton-Smith, eds., The Study of Games (New York: Wiley, 1971); S. Culin, "Mankala, the National Game of Africa," in Annual Report of the U.S. National Museum (Washington: Government Printing Office, pp. 597-606. Also in The Study of Games, pp. 94-102.
47. See Henri Maurier, "Do we have an African Philosophy?" and Bodunrin, "The Question of African Philosophy," in Wright, African Philosophy, pp. 25-40 and 1-24 respectively.
48. Bodunrin, "The Question of African Philosophy," Philosophy, Vol. 56, 1981, pp. 161-79; Odera Oruka, "The Fundamental

Principles in the Question of 'African Philosophy' - I,"
Second Order, Vol. IV, No. 1, 1975, pp. 44-45.
49. This point was well emphasized in Professor George
Chatalian's Inaugural Lecture. See our note 13 above.
50. Chatalian, Inaugural Lecture, Part V, section 5. Professor
Chatalian's full discussion on this issue is contained in
Parts IV and V of his Inaugural Lecture.
51. Philosophers in the 1980s are getting more involved in
subjects that could hardly have been admitted into
philosophy programs two decades ago. In the United States
topics that are relevant to societal needs are now
considered such as feminism, social work, environmental
pollution, literature, medicine, medical ethics, being
human in a technological age, science, as well as
technology and human values.
52. Jean-Paul Lebeuf's speech was originally quoted from
Presence Africaine, Vol. 53, 1965, p. 129 by W. A. Hart,
"The Philosopher's Interest in African Thought: A
Synopsis," _Second Order_, Vol. I, No. 1, 1972, pp. 43-44.
But the quotation as presented above is from Wright,
"Investigating African Thought," in Wright, _African
Philosophy_, p. 43.
53. See M. Akin Makinde, "A Philosophical Analysis of the
Yoruba Concepts of Ori and Human Destiny" and "An African
Concept of Human Personality: The Yoruba Example,"
International Studies in Philosophy, Vol. XVII, No. 1,
1985, pp. 53-69, and _Ultimate Reality_ and Meaning, Vol. 7,
No. 3, 1984, pp. 189-200, respectively. Professor Kwasi
Wiredu has also given a sound philosophical analysis of the
Akan notion of Truth. See his recent book, _Philosophy and
African Cultures_ (Cambridge: Cambridge University Press,
1980), Part III.
54. Wright, "Investigating African Philosophy," p. 53.
55. See Makinde, "Immortality of the Soul."
56. Onwuanibe, "The Human Person and Immortality," in Wright,
African Philosophy, p. 183.
57. Onwuanibe, "The Human Person and Immortality," pp. 183,
185.
58. See Makinde, "An African Concept of Human Personality: The
Yoruba Example."
59. Wande Abimbola, _Ifa: An Exposition of Ida Literary Corpus_
(Ibadan: Oxford University Press, 1976), pp. 113-49. See
also Abimbola, _La Notion_.
60. Abimbola, _Ifa Divination Poetry_, pp. 117, 148, and Makinde,
"A Philosophical Analysis."
61. Gilbert Ryle, _Dilemmas_ (Cambridge: Cambridge University
Press, 1966), Ch. 2. See also Makinde, "A Philosophical
Analysis" above.
62. Makinde, "An African Concept," and Gyekye, "Akan Concept of
a Person."

63. Makinde, "Ifa as a Repository of Knowledge," and Wande Abimbola, "Ifa as a Body of Knowledge and as an Academic Discipline," in Lagos Notes and Records, Vol. I, No. 1, June 1967.
64. See Sodipo, "Notes on the Concept of Cause" and Helaine K. Minkus, "Causal Theory," in Wright, African Philosophy, pp. 113-48; Henry Olela, "The African Foundations of Greek Philosophy," in Wright, African Philosophy, pp. 77-92; John A. A. Ayoade, "Time in Yoruba Thought," in Wright, African Philosophy, pp. 92-112; Wiredu, Philosophy and African Culture, Part II; Onwuanibe, "The Human Person and Immortality," in Wright, African Philosophy, and Gyekye, "Akan Concept of a Person," in Wright, African Philosophy.
65. Thomas J. Blakeley, "The Categories of Mtu and The Categories of Aristotle," and Benjamin Eruku Oguah, "African and Western Philosophy: A Comparative Study," in Wright, African Philosophy, pp. 163-70 and 213-26 respectively.
66. See Obafemi Awolowo, The People's Republic (Ibadan: Oxford University Press, 1968), Chapter 9, and his Thoughts on Nigerian Constitution (Ibadan, Oxford University Press, 1966). Both works deal with Awolowo's idea of "The Regime of Mental Magnitude." For a good analysis of Awolowo on mental magnitude, see M. Akin Makinde, "'Mental Magnitude': Awolowo's Search for Ultimate Reality, Meaning, and the Supreme Value of Human Existence," Ultimate Reality and Meaning, Vol. 10, No. 1, March 1987, pp. 3-13.
67. See Julius Nyerere, Freedom and Socialism: Uhuru na Ujamaa (Dar-e-Salaam: Oxford University Press, 1968); Julius Nyerere, Ujamaa: Essays on Socialism (London: Oxford University Press, 1971), and Julius Nyerere, Man and Development (London: Oxford University Press, 1974); Leopold Sedar Senghor, On African Socialism (New York: Praeger, 1964), and his Liberté (Paris: Editions de Seuil, 1964); Kwame Nkrumah, Consciencism: Philosophy and Ideology (New York: Monthly Review Press, 1964), and his I Speak of Freedom: A Statement of African Ideology (New York: Praeger, 1961).
68. Bodunrin, "The Question of African Philosophy. Some African philosophers have seen ideology as philosophy. See Wiredu, "Philosophy and African Culture," in Wright, African Philosophy, Part II, Section 5 on "Marxism, Philosophy and Ideology."
69. See note 52 above.
70. The University of Hawaii, Honolulu, recently established the Institute for Comparative Philosophy. This institute has organized a program whose aim is to provide an intensive training dealing with the incorporation of Asian and comparative philosophical materials into standard undergraduate and graduate courses in philosophy. There is

the hope that the Institute's program will include African philosophy.

71. Paulin Hountondji seems to have made the same point when he proposed that only native Africans are able to correctly study, analyze, and interpret the thought of African people. See Wright, African Philosophy, p. 51. Wright's source of information is from Hountondji's paper, "Comments on Contemporary African Philosophy," Diogenes, Vol. 71, 1970, pp. 109-30.

72. By "unspoken" and "unwritten" language of African philosophy I mean the original or native langauge in which ordinary Africans think but which are never spoken and written when African philosophy is done, written, or taught by African philosophers.

73. Hountondji, "Comments on Contemporary African Philosophy," p. 109.

Chapter 4

1. Michael A. Snyder, "Is a New Dark Age Coming?" The Plain Truth, January 1984, p. 40.
2. Abimbola, La Notion, pp. 75-76.
3. Abimbola, La Notion, p. 76.
4. Nyerere, Ujamaa, p. 9.
5. Nyerere, Ujamaa, p. 12. Thus the first article of TANU's creed is, "Binadamu wote ni nduga zangu, na Afrika ni moja" (I believe in Human Brotherhood and the Unity of Africa. 'Ujamaa' then, or 'Familyhood,' describes our socialism.) Senghor, On African Socialism, p. 94.
6. Nkrumah, Consciencism: Philosophy, p. 77. Awolowo, The People's Republic, p. 208.
7. Nyerere, Ujamaa, p. 11.
8. Senghor, On African Socialism, p. 59.
9. Nkrumah, Consciencism, p. 78.
10. Senghor, On African Socialism, p. 165.
11. Hountondji, African Philosophy, Part Two, Sections 6 & 7.
12. Awolowo, Peoples Republic, pp. 190-91.
13. Obafemi Awolowo, Text of a Public Lecture delivered at the University of Ife, Ile-Ife, Nigeria on 9 April 1970, as the Chancellor of the University (Lagos: Ibadan University Press, 1970), p. 6.
14. Awolowo, Text, p. 6.
15. Awolowo, People's Republic, p. 209.
16. Awolowo, People's Republic, p. 209.
17. Awolowo, People's Republic, pp. 209-10.
18. Makinde, "Technology Transfer."
19. Awolowo, People's Republic, p. 210.
20. Awolowo, People's Republic, p. 210.
21. Omoregbe Nwanwene, "Awolowo's Political Philosophy," in Quarterly Journal of Administration (Institute of

Administration, University of Ife), Vol. IV, No. 2, 1970, p. 129. See also Obafemi Awolowo, Voice of Reason (Akure: Fagbamigbe Publishers, 1981), pp. 196-97.

22. Nwanwene, "Awolowo's Political Philosophy," p. 129. Obafemi Awolowo, Path to Nigerian Freedom (London: Faber, 1947), p. 25.

23. Awolowo, People's Republic, p. 75.

24. Awolowo, People's Republic, pp. 188-89; 163-65ff.

25. Nkrumah, Consciencism, p. 77.

26. Awolowo, People's Republic, pp. 191-92.

27. See R. Italiander, The New Leaders of Africa (London, 1961), pp. 278-79.

28. Awolowo, People's Republic, pp. 191ff. See also Obafemi Awolowo, The Problems of Africa: The Need for Ideological Re-appraisal (London: Macmillan, 1977), pp. 63-64.

29. See Obafemi Awolowo, Voice of Wisdom (Akure: Fagbamigbe Publishers, 1981), pp. 39-46.

30. In fact, the official manifesto issued by the Action Group of Nigeria on the eve of Nigeria's Independence in 1960 is entitled: "Democratic Socialism." See Obafemi Awolowo, "Case for Ideological Orientation," in Voice of Reason, pp. 184-95, especially p. 186.

31. This was the official motto of the Action Group, the political party led by Awolowo. See Owolowo, Voice of Reason, pp. 196; 166-76.

32. Awolowo, People's Republic, p. 206.

33. Akin Omoboriowo, Awoism: A Select Theme of The Complex Ideology of Chief Obafemi Awolowo (Ibadan: Evans Brothers, 1982), p. 19. See Marx and Engels, The Communist Manifesto, in Robert C. Tucker, ed., The Marx-Engels Reader, 2nd edition, New York: W. W. Norton & Company, 1978, p. 489.

34. Awolowo, People's Republic, p. 206. See also Awolowo, Voice of Reason, pp. 177-84.

35. Awolowo, People's Republic, p. 192.

36. Nkrumah, Consciencism, p. 77. See also James Anquandah, Together We Sow and Reap (Accra: Assempa Publishers, 1979), pp. 115ff.

37. Nyerere, Freedom and Socialism, p. 13.

38. Obafemi Awolowo, Thoughts on Nigerian Constitution (Ibadan: Oxford University Press, 1966), Chapter VI. For a good commentary on Awolowo's doctrine, see Omoboriowo's Awoism, Chapter 3.

39. See note 25 on Nyerere above. Awolowo's view is more elaborately stated in his doctrine of mental magnitude.

40. Awolowo, Problems of Africa, p. 53.

41. Awolowo, Problems of Africa, p. 54.

42. Awolowo, People's Republic, p. 211.

43. Awolowo, People's Republic, p. 213.

44. Awolowo, People's Republic, p. 214.

45. Awolowo, People's Republic, p. 206, 228-29. Awolowo also sees thinking as a spiritual process, and "the only way to exercise the mind is constantly to engage in clear, decisive, calm, deliberate, sustained, and constructive thinking with a definite end in view, which end should benefit the thinker as well as others (p. 227).
46. Awolowo, People's Republic, p. 227.
47. Awolowo, People's Republic, p. 226.
48. John Stuart Mill, A System of Logic: Ratiocinative and Inductive (London: Longman, 1970), 8th edition, Book VI. See also J. M. Robson, The Improvement of Mankind (Toronto: University of Toronto Press, 1968).
49. Awolowo, People's Republic, p. 227.
50. Awolowo, People's Republic, p. 227.
51. Awolowo, Thoughts, pp. 157-58.
52. Awolowo, Thoughts, p. 158.
53. Awolowo, Thoughts, p. 158.
54. Awolowo, People's Republic, p. 268.
55. Awolowo, People's Republic, p. 215-16.
56. Awolowo, People's Republic, p. 215.
57. Obafemi Awolowo, Path to Nigerian Greatness (Enugu: Fourth Dimension Publishing Co., 1981), pp. 149-59.
58. Awolowo, Thoughts, p. 159.
59. Awolowo, People's Republic, p. 188.
60. Awolowo, People's Republic, p. 206, 229.
61. H. A. Oluwasanmi, "Foreword" to Omoboriowo, Awoism, p. xi.
62. Obafemi Awolowo, Path to Nigerian Greatness, Part III, Chapter 12, pp. 135-48.
63. Omoboriowo, Awoism, p. xvi.
64. B. A. Ogundimu, "Personality Variable in Political Leadership and Decision-Making: An Analysis of Obafemi Awolowo's Operational Codes," Quarterly Journal of Administration, Vol. XII, No. 3, April 1978, p. 237.
65. Awolowo, Thoughts, p. 159.
66. John Stuart Mill, Utilitarianism (Indianapolis: Bobbs-Merrill, 1957), p. 14. The quotation from Mill is used in connection with his argument that human beings have faculties more elevated than the animal appetites, and his conviction about the superiority of mental over bodily pleasures because of the greater permanency and safety of the former (pp. 11-5).
67. See Plato, Apology. Socrates lived circa 470-399 B.C.
68. Nagel, "Defense," in Edwards and Pap, eds., A Modern Introduction to Philosophy, p. 461.
69. Nagel, "Defense," in Edwards and Pap, eds., A Modern Introduction to Philosophy, p. 471. (Italics mine.)
70. Omoregbe Nwanwene, "Awolowo's Strategy and Tactics of the People's Republic of Nigeria - A Review Article," Quarterly Journal of Administration, Vol. V, No. 2, January 1971, p. 229.

71. Nagel, "Defense," pp. 462-63. I have criticized Nagel's position in my article titled "Pascal's Wager and the Atheist's Dilemma," *International Journal for Philosophy of Religion*, Vol. 17, No. 3, 1985, pp. 115-129, with particular reference to Awolowo, pp. 123-124.
72. Quotation from Ian Harvey, *The Techniques of Persuasion: An Essay in Human Relationship* (London: Falcon Press, 1951), p. 2.
73. George Orwell, *1984* (New York: New American Library, 1983), pp. 151-245. George Orwell's real name was Eric Blair, and the name "Orwell" is the name of a river in England.
74. Orwell, *1984*, pp. 170-71.
75. Orwell, *1984*, pp. 171, 175-77.
76. M. Akin Makinde, "God and Social Reality: Reflections on the Pains of Growth," a paper presented at the New Ecumenical Association Conference at Dorado Beach, Puerto Rico, 30 December 1983 to 4 January 1984.
77. Quotation from Auden is taken from Jonathan Schell, *The Fate of the Earth* (New York: Avon Books, 1982), p. 230.
78. Jack A. Krieger, "Manhood and Brotherhood," *The Christian Science Journal*, October, 1974, p. 592.
79. Gilkey, "The Religious Dilemmas," p. 86.
80. Awolowo, *People's Republic*, pp. 205, 211, 231.
81. Awolowo, *People's Republic*, p. 214.
82. The Fabian Society, founded in 1883, applied Mill's ideas to morals, society, politics, and economics. See Adam B. Ulam, *Philosophical Foundation of English Socialism* (New York: Octagon Books, 1964), p. 678, and A. M. McBriar, *Fabian Socialism and English Politics, 1884-1918* (Cambridge: Cambridge University Press, 1962), p. 7. See also Ruth Borchard, *John Stuart Mill: The Man* (London: C. A. Watts, 1957), p. 147 and D. C. Somervell, *English Thought in the Nineteenth Century* (London: Methuen, 1964), p. 96. See also Harold Wilson, *The Relevance of British Socialism* (London: Weindenfield & Nicholson, 1964), p. 9.
83. Awolowo, *Problems of Africa*, p. 69.
84. Omoboriowo, *Awoism*, pp. 50, 11.

Chapter 5

1. Huard, "Western Medicine," pp. 211-12.
2. Huard, "Western Medicine," p. 213.
3. Quoted from my paper, "Cultural and Philosophical Dimensions of Neuro-Medical Sciences," delivered at the "1982 Joint Conference of the Association of Psychiatrists in Nigeria, the African Psychiatric Association and the World Federation for Mental Health," held at the University of Ife, 20-23 September 1982. Published in *Nigerian Journal of Psychiatry*, September 1987, 85-100.

4. Personal communication with Ifayemi Eleburuibon, an Ifa priest from Osogbo, Nigeria. See also Wande Abimbola, "Ifa as a Body of Knowledge and as an Academic Discipline," in Lagos Notes and Records, pp. 30-34.
5. Ifayemi Eleburuibon, note 4 above. See also Adewale Thompson, African Beliefs: Science of Superstition? pp. 267-68.
6. Bombastus Paracelsus, astrologer, alchemist and miracle worker, treats in detail the conception of "magus" under his discussion on the "Fundamentals of Magic Medicine," in his book Sagacious Philosophy (Philosophia Sagax), as discussed in H. M. Patcher, Paracelsus: Magic into Science, pp. 78ff.
7. Ifayemi Eleburuibon, note 4 above.
8. Personal communication, Chief J. A. Lambo, President of the Nigerian Association of Medical Herbalists.
9. Personal communication, note 8 above and from Chief J. A. Abiola, the Eisikin of Aiyegunle-Ekiti, Ondo State of Nigeria. A great deal of my paper on "Cultural and Philosophical Dimensions of Neuro-Medical Sciences," note 3 above, is devoted to the treatment of functional illness in Yoruba society.
10. Huard, "Western Medicine," pp. 213-15.
11. Abimbola, La Notion, p. 77.
12. Personal communication from Daniel Kayode Makinde, a well-known traditional header.
13. See note 12 above.
14. In Yoruba society this is sometimes regarded as a cause of some illness--organic or functional. It could be an occult cause like epe (curse) or ase/afose whose powerful incantary words or verses have magical effects on the targeted victims. This category of cause is usually attributed to evil men (Aiyekunrin). The second kind of possible causes are spiritual causes, usually attributed to the witches. Spiritual causes, like witches, are usually attributed to evil women (Aiyebirin). These are awon aje or eleye (witches). Their powers are said to be in their private parts. A detailed analysis of this occurs in my paper "Cultural and Philosophical dimensions of Neuro-Medical Sciences," note 3 above.
15. Huard, "Western Medicine," p. 216.
16. See Thompson African Beliefs, Chapters 7 and 8.
17. See Dr. A. O. Sanda, "The Scientific or Magical Ways of Knowing: Implications for the Study of African Traditional Healers," Second Order, Vol. VII, Nos. 1 & 2, January and July, 1978, pp. 70-83. For a related discussion see Bronislaw Malinowski, Magic, Science and Religion and Other Essays (New York: Doubleday, 1954).
18. This is my own adaptation from J. J. C. Smart, Between Science and Philosophy (New York: Random House, 1968), p. 16.

19. Personal communication from Chief James Abiola, the Eisikin of Aiyegunle-Ekiti, Nigeria.
20. Personal communication, Daniel K. Makinde, Chief J. A. Abiola, and Chief J. A. Lambo. The power of the words spoken in incantation as related to the nature of these animals was later explained to me by the last two informants.
21. Thompson, African Beliefs, p. 131.
22. Thompson, African Beliefs, p. 132-34.
23. Thompson, African Beliefs, p. 135, 137. Extract of Omoleye's speech from The Sunday Sketch of March 5 & 12, 1978.
24. Dr. Farrow was quoted in Omoleye's speech The Sunday Sketch of March 5 & 12, 1978.
25. Omoleye, note 24 above. The reality of the power of African traditional science and medicine has also been discussed in Ezeabazili Nwankwo, African Science; Myth or Reality? (New York: Vantage Press, 1978). This book provides testimonies of what the author describes as "the wonders of African science."
26. What may be regarded as the bible of the school of General Semantics in Chicago is A. Korzybski, Science and Sanity (Lakeville, CT: International Non-Aristotelian Library, 1948). Korzybski, however, drew his inspiration from the philosophical work of C. K. Ogden and I. A. Richards, Meaning of Meaning (New York: Harcourt, Brace and World, Inc., 1923). Ogden's and Richards' work was a major contribution to the philosophical tradition which considers the study of meaning to have an important bearing on human happiness and sanity. The effects of words used in some dreaded African medicine like Ase, Olugbohun, Ayajo, and Gbetagbetu are good cases that can be studied as they definitely have wonderful effects on human beings and situations. A good and general discussion on "General Semantics" can be found in Anatol Rapoport, Operational Philosophy (San Francisco: International Society for General Semantics, 1969), Chapter 18.
27. Thompson, African Beliefs, p. 264.
28. Thompson, African Beliefs, p. 264.
29. I witnessed many of these deliveries from my father who was responsible for the deliveries of about ninety-five percent of babies born at Imesi-Ekiti between 1939-1981. He died in 1981 at the age of 77. The mid-wives at the dispensary in this town had occasions to call on him to deliver still-born babies that would normally have required Caesarian operations. All these were done successfully. The "surgical" tools were his hands and powerful incantations with a special kind of ase known as olugbohun. Thompson describes it as "a powerful charm believed to be a representation of the echo and is reputed to act as a catalytic

force to words of power which the ancient mystics have labored so hard to find" (p. 138).

30. I use the term "afrotherapy" to cover the most common kind of the Yoruba traditional treatment of mental illness. See my paper "Cultural and Philosophical Dimensions of Neuro-Medical sciences," note 3 above.

31. T. Adeoye Lambo, "Traditional African Cultures and Western Medicine," in Poynter, Medicine and Culture, p. 203. See also E. H. Ackerknecht, "Problems of Primitive Medicine," Bulletin of the History of Medicine, Vol. II, 1942, pp. 501-21.

32. Lambo, "Traditional African Cultures," pp. 204-5.

33. Lambo, "Traditional African Cultures," p. 209. Professor Lambo also refers to what he calls Turner's "Classic study of divination and diagnosis of diseases among the Ndembu tribe of Central Africa" which, he thinks, "has demonstrated how in this diagnostic exercise, the Ndembu diviner not only refers to the influences of unseen spirits, but also emphasizes that the patient's condition is due to a whole series of upsets and disquiet in his social field." Turner referred to divination as "social analysis." He observed that the Ndembu believe that a patient will not recover from his mental illness "until all the tensions and aggressions in the group's interrelations have been brought to light and exposed to ritual treatment." Reporting his own experience as a psychiatrist who has combined the Western with traditional method of healing Lambo points out that "much of what is done in Yoruba diagnostic protocols resemble the well-known projective techniques of modern psychology" (Lambo, "Traditional African Cultures," p. 209). For Lambo's reference to Turner, see V. W. Turner, Ndembu Divination, Its Symbolism and Techniques (Manchester: Manchester University Press, 1961), Rhodes-Livingston Papers, No. 31.

34. Abayomi Sofowora, "Man, Plants, and Medicine in Africa," an Inaugural lecture delivered at the University of Ife, Ile-Ife, Nigeria on 7 January 1981, p. 5.

35. Note 34 above, pp. 6, 7, 20-22. Sofowora has reported that evidence from chemo-taxonomic studies has supported the idea which led to the change of the genus fagara to Zanthoxyloides or Zanthoxylum Zanthoxyloides, for the treatment of sickle cell anemia. See J. O. Moody, Phylochemical examination of Zanthoxylum (fagara) rubescenes, M. Phil. Thesis, University of Ife, 1980. Requests for patents of this drug has been received from many multinational pharmaceutical companies.

36. See J. A. Lambo, "Integration of Traditional Medicine: Another View," Nigerian Tribune, 11 August 1983, p. 2.

37. See Huard, "Western Medicine," p. 217.

38. Chief J. A. Lambo whose special field in traditional medicine is in obstetrics and gynecology used the word

136

"synthetic" to describe some aspects of Western medicine. His contention is that the herbalist or native doctor relies on the use of pure, natural herbs, roots and animal substances for drugs as opposed to the synthetically manufactured drugs used in Western medicine.

39. After preparation (which always includes a small portion of the earth which contains an early urine of the pregnant woman, and tied with white threads in an empty shell of alligator pepper), the medicine is nailed to the wall. It is not removed until the ninth month of pregnancy. And unless it is removed, the woman does not go into labor; the woman may carry the pregnancy for as long as the medicine remains nailed to the wall. It is therefore advisable that the women knows the location of the medicine just in case the herbalist dies before delivery time.

40. Such cases include still-born babies, breech babies, placenta previa, some of which normally require Caesarean operations in modern hospitals. Complicated cases were usually referred either directly from the individual patient or from the local dispensary. No fees were charged for each delivery. An individual gave what she could, ranging from a bottle of local wine to a live chicken.

41. Sofowora, note 34 above, p. 14. See "Traditional Medicine in Zaire: Present and Potential Contribution to the Health Services," (Ottawa, Ontario: International Development Research Center, 1980), and the Bulletin de Medecine Traditionelle au Zaire et en Afrique, a publication of Zaire's Healers Medicare Centre.

42. Sofowora, note 34 above, pp. 14-15.

43. See N. H. Keswani, "Modern Medicine in a Traditional India Setting," in Poynter, Medicine and Culture, p. 191, and Joseph Needham and Lu Gwei-Djen, "Chinese Medicine," in Poynter, Medicine and Culture, p. 288.

44. Keswani, "Modern Medicine," p. 193.

45. Noel Poynter, Medicine and Man (Harmondsworth, Middlesex: Penguin Books, 1971), p. 26.

46. Chief J. A. Lambo, as reported by Sina Adedipe, "Native Doctors Go Modern," Sunday Concord, 10 July 1983, p. XI.

47. See Lambo, "Integration," note 36 above. The paper was a response to the controversy in Nigeria, between Western trained doctors and traditional physicians over the important issue of recognition and integration of traditional with Western medicine. Chief Lambo was highly critical of a particularly "hypocritical and one-sided" opinion by Dr. Sam Nwangoro in the Nigeria Daily Times of 19 July 1983.

48. According to Chief Lambo, note 36 above, p. 2,

We traditional doctors admire the knowledge and the professional ability of modern doctors, especially in the field of surgery, diagnosis, sedative medicine, and anaesthesia, etc., hence many of us are desirous that our children should combine the two as practised in some civilized countries. I assert with no fear of contradiction that some private doctors in this country who combine the two get more to do because our people always refer cases to them as doctors who know how to 'change hands.' The traditional healers cure such dreaded diseases as cancer and diabetes and others like jaundice, asthma, ulcer, etc., with herbal drugs which Nigerian modern doctors take to be a cure by chance. The use of herbal drugs for the cure of the above and other diseases had been in existence with our forefathers and well before the introduction of modern medicine into Nigeria.

Therefore, Chief Lambo could not see the reason why the Nigerian medical doctors did not encourage the integration through research, of traditional with modern medicine as well as the building of hospitals for the practice of integrated medicine.

49. Benjamin Walker, Encyclopedia of Metaphysical Medicine (London: Routledge & Kegan Paul, 1978), p. 224.
50. Pierre Huard, "Western Medicine," p. 216.
51. Popper, Objective Knowledge, pp. 348, 84. Popper sees science as the growth of knowledge through criticism and inventiveness (p. 86) and the "task which science sets itself and the main ideas which it uses are taken over without any break from prescientific mythmaking," p. 348.
52. T. A. Lambo, "Traditional African Cultures," p. 201.

Chapter 6

1. Maurier, "Do We Have an African Philosophy?" in Wright, African Philosophy, pp. 25-40; H. Odera Oruka, "The Fundamental Principles in the Question of African Philosophy," Second Order, Vol. IV, 1975; and Bodunrin, "The Question of African Philosophy," in Wright, African Philosophy, pp. 1-23.
2. All the above constitute part of the contributions to the debate. Others are R. E. Ruch, "Is There an African Philosophy?" Second Order, Vol. III, No. 2, 1974; Kwasi Wiredu, "On an African Orientation in Philosophy," Second Order, Vol. I, No. 2, 1972; Paulin Hountondji, "African Philosophy: Myth and Reality," Thought and Practice, Vol. 1, 1974; and lately, P. O. Bodunrin, "The Question of African Philosophy," Philosophy, Vol. 56, 1981, above. The first edition of Richard Wright's book appeared in 1977.
3. See Makinde, "Technology Transfer."

4. This is not unusual in Africa. In Nigeria, for instance, Professor G. A. Makanjuola in the Department of Agricultural Engineering in the Faculty of Technology, University of Ife, invented a pounding machine for one of the national foods, pounded yam. Neither the Federal Government nor the University took the matter seriously until the diagram somehow got into the hands of a Japanese firm and, before we knew where we were, the Nigerian market was flooded with yam pounding machines as a new addition to the list of technological exports to Nigeria. Quite recently, Professor O. Odeyemi, Head of Department of Microbiology at the same University discovered the utilization of refuse for the production of methane gas. The discovery was reported on a National Television Service (Ibadan) but that was the end of the matter.

LITERATURE CITED

Abimbola, Wande. Ifa Divination Poetry. New York: Nok
Publications, 1977.
_____. Awon Oju Odu Mereerindinlogun. Ibadan: Oxford
University Press, 1977.
_____. Ifa: An Exposition of Ifa Literary Corpus. Ibadan:
Oxford University Press, 1976.
_____. Sixteen Great Poems of Ifa. Niamey: UNESCO, 1975.
_____. La Notion de Personne en Afrique Noire. Paris: Centre
National de la Recherche Scientifique, 1971.
_____. Ijinle Ohun Enu Ifa, "Apa Keji," Vol. 2, Glasgow:
Collins, 1969.
_____. Ijinle Ohun Enu Ifa, "Apa Kini," Vol. 1, Glasgow:
Collins, 1968.
_____. "Ifa as a Body of Knowledge and as an Academic
Discipline." Lagos Notes and Records, I, no. 1 (June
1967).
Ackerknecht, E. H. "Problems of Primitive Medicine." Bulletin
of the History of Medicine, II (1942):501-21.
Adedipe, Sina. "Native Doctors Go Modern." Sunday Concord, 10
July 1983, p. XI.
Alston William P., and Brandt, Richard B. The Problems of
Philosophy (Introduction). Boston: Allyn & Bacon, 1967.
Anquandah, James. Together We Sow and Reap. Accra: Assempa
Publishers, 1979.
Avedon E. M., and Sutton-Smith, B., eds. The Study of Games.
New York: Wiley, 1971.
Awolowo, Obafemi. Voice of Reason. Akure: Fagbamigbe
Publishers, 1981.
_____. Path to Nigerian Greatness. Enugu: Fourth Dimension
Publishing Co., 1981.
_____. The Problems of Africa: The Need for Ideological Re-
appraisal. London: Macmillan, 1977.
_____. Text of a Public Lecture delivered at the University
of Ife, Ile-Ife, Nigeria on 9 April 1970, as the Chancellor
of the University. Lagos: Ibadan University Press, 1970.
_____. The People's Republic. Ibadan: Oxford University
Press, 1968.
_____. Thoughts on Nigerian Constitution. Ibadan: Oxford
University Press, 1966.
_____. Path to Nigerian Freedom. London: Faber, 1947.

Ayer, A. J. Language, Truth and Logic. 18th ed. London: Victor Gollanz, 1970.

Ayoade, John A. A. "Time in Yoruba Thought," in Richard A. Wright, ed., African Philosophy. 3rd ed. New York: University of America Press, 1983.

Basalla, George, Coleman, William, and Kargon, Robert H., eds. Victorian Science: A self-portrait from the Presidential Address of the British Association for the Advancement of Science. New York: Doubleday, 1970.

Blakeley, Thomas J. "The Categories of Mtu and The Categories of Aristotle," in Richard A. Wright, ed., African Philosophy. 3rd ed. New York: University of America Press, 1983.

Bodunrin, P. O. "The Question of African Philosophy," in Richard A. Wright, ed., African Philosophy. 3rd ed. New York: University of America Press, 1983. Reprinted from Philosophy (1981):161-79.

Borchard, Ruth. John Stuart Mill: The Man. London: C. A. Watts, 1957.

Burke, John G. ed., The New Technology and Human Values. Belmont, CA: Wadsworth, 1972.

Carnap, Rudolf. "Elimination of Metaphysics through Logical Analysis of Language," in A. J. Ayer, ed., Logical Positivism. New York: The Free Press, 1959.

Caws, Peter. Philosophy of Science. Princeton: Van Nostrand, 1966.

Chatalian, George. "Philosophy, The World and Man: A Global Conception." Inaugural Lecture delivered at the University of Ife, Ile-Ife, Nigeria, on 28 June 1983.

Chokr, Nader. "Mankala: Wari & Solo: A Study of a Board Game Played by African People." Archives of Smithsonian Institution, Washington, D.C., 1985.

Collingwood, R. G. Quote in Lionell Rubinoff, Collingwood and the Reform of Metaphysics. Toronto: University of Toronto Press, 1970.

_____. Speculum Mentis. Oxford: Clarendon Press, 1924.

Culin, S. "Mankala, the National Game of Africa," in E. M. Avedon and B. Sutton Smith, eds., The Study of Games. New York: Wiley & Sons, 1971, 94-102.

De Wulf, Maurice. Scholastic Philosophy: Medieval and Modern, translation by Peter Coffey. New York: Dover, 1956.

Engels, Frederick, and Karl Marx. The Communist Manifesto, in Robert C. Tucker, ed., The Marx-Engels Reader. 2nd edition. New York: W. W. Norton, 1978, pp. 469-500.

Ewing, A. C. "Meaninglessness." in Paul Edwards and Arthur Pap, eds., A Modern Introduction to Philosophy. New York: The Free Press, 1965.

Feibleman, James K. Understanding Philosophy. New York: Horizon Press 1973.

Fichte, John Gottlieb. Quote in Theodor Oizerman, Problems of the History of Philosophy. Translated by Robert Daglish. Moscow: Progress Publishers, 1973.

142

Finazz, A. "Una Filosofia Africana?" Africa (Rome), 29 (1974).
Gilkey, Langdon. "The Religious Dilemmas of a Scientific Culture: The Interface of Technology, History, and Religions," in Donald M. Borchert and David Stewart, eds., Being Human in a Technological Age. Athens: Ohio University Press, 1979.
Gilson, E. The Christian Philosophy of St. Augustine, translation by L. E. M. Lynch. London, 1961.
Gilson, H. I. St. Augustine and His Influence Through the Ages. London, 1958.
Gyekye, Kwame. "Akan Concept of a Person," in Richard A. Wright, ed., African Philosophy. 3rd ed. New York: University of America Press, 1983.
_____. "African Religions and Philosophy," a review article on John Mbiti, Second Order, IV, no. 1, (1975):86-94.
Hallen, Barry. "Robin Horton on Critical Philosophy and Traditional Thought." Second Order, VI, no. 1 (1977):81-92.
Hart, W. A. "The Philosopher's Interest in African Thought: A Synopsis." Second Order, I, no. 1 (1972).
Harvey, Ian. The Techniques of Persuasion: An Essay in Human Relationship. London: Falcon Press, 1951.
Horton, Robin. "Traditional Thought and the Emerging African Philosophy Department: A Comment on the Current Debate." Second Order, VI, no. 1 (1977):64-80.
_____. "African Traditional Thought and Western Science," in Africa, XXXVII, nos. 1 & 2, (1967).
_____. "The Kalabari World-View: An Outline and Interpretation." Africa, XXXII, no. 3 (1962):197-220.
Hountondji, Paulin. African Philosophy: Myth and Reality. Bloomington: Indiana University Press, 1983.
_____. "African Philosophy: Myth and Reality?" Thought and Practice, 1 (1974).
_____. "Comments on Contemporary African Philosophy." Diogenes, 71 (1970):109-30.
_____. Interviewed. West Africa, (22 August 1983):1955.
Huard, Pierre. "Western Medicine and Afro-American Ethnic Medicine" in F. N. L. Poynter, ed., Medicine and Culture. New Series, Vol. XV. London: Wellcome Institute of the History of Medicine, 1969.
Hubble, Douglas. "Medicine and Culture" in F. N. L. Poynter, ed., Medicine and Culture. London: Wellcome Institute of the History of Medicine, 1969.
Hume, David. A Treatise of Human Nature. Selby-Bigge edition, Oxford: Clarendon Press, 1967.
_____. Enquiry Concerning the Human Understanding. Selby-Bigge edition. Oxford: Clarendon Press, 1962.
Hyman, Arthur and Walsh, James J. Philosophy in the Middle Ages. New York: Harper & Row, 1967.
Italiander, R. The New Leaders of Africa. London, 1961.
Jahn, Janheinz. Muntu. New York: Grove Press, 1961.

Jaspers, Karl. "Socrates, Buddha, Confucius, Jesus: The Paradigmatic Individuals," in Hannah Arendt, ed., The Great Philosophers. Vol. 1. Translated by Ralph Manheim. New York: Harcourt, Brace & World, 1962.

Jevons, William Stanley. The Principles of Science: A Treatise on Logic and Scientific Method. New York: Dover, 1958.

Joad, C. M. Philosophy. London: Richard Clay, 1974.

Kagame, Alexis. Rwandan-Bantu Philosophy of Being. Brussells, 1956.

Keita, Lucinay. "The African Philosophical Tradition," in Richard A. Wright, ed., African Philosophy. 3rd ed. New York: University of America Press, 1983.

Keswani, N. H. "Modern Medicine in a Traditional India Setting," in F. N. L. Poynter, ed., Medicine and Culture. London: Wellcome Institute of the History of Medicine, 1969.

Korzybski, A. Science and Sanity. Lakeville, CT: International Non-Aristotelian Library, 1948.

Krieger, Jack A. "Manhood and Brotherhood." The Christian Science Journal (October, 1974).

Lambo, J. A. "Integration of Traditional Medicine: Another View." Nigerian Tribune, 11 August 1983, p. 2.

Lambo, T. Adeoye. "Traditional African Cultures and Western Medicine," in F. N. L. Poynter, ed., Medicine and Culture. London: Wellcome Institute of the History of Medicine, 1969.

Laplace, Pierre Simon. Essai Philosophique (1814) which has been published in English under the title A Philosophical Essay on Probabilities (New York: Dover Publications, 1952).

_____. Theory of Probability (Introduction). Translation from the first French edition (1812) under the title Theorie Analytique des Probabilities.

Levy-Bruhl, Lucian. Notebooks on Primitive Mentality, New York: Harper & Row, 1975.

_____. Primitive Mentality. Boston: Beacon Press, 1966, reprint of 1923 edition.

_____. Now Natives Think. New York: Alfred Knopf, 1925.

Locke, John. An Essay Concerning Human Understanding. Book 4. Oxford: Clarendon Press, 1969.

Makinde, M. Akin. "The Possibility of an African Continental Language: A Philosophical Analysis," Journal of African Studies (UCLA), forthcoming.

_____. "'Mental Magnitude': Awolowo's Search for Ultimate Reality, Meaning, and Supreme Value of Human Existence," Ultimate Reality and Meaning, 10, no. 1 (1987):3-13.

_____. "Cultural and Philosophical Dimensions of Neuro-Medical Sciences." Delivered at the 1982 Joint Conference of the Association of Psychiatrists in Nigeria, the African Psychiatric Association and the World Federation for Mental Health, held at the University of Ife, 20-23 September 1982. Also in Nigerian Journal of Psychiatry (September 1987):85-100.

_____. "Technology Transfer: An African Dilemma" in John W. Murphy, Algis Mickunas and Joseph J. Pilotta, eds., The Underside of High Technology, New York/London: Greenwood Press, 1986, 177-189. A revised version of paper delivered to the College of Engineering, Ohio University, Athens, Ohio, 15 May 1984.

_____. "The World and its Enemies: A Philosophical Perspective." Presented at "The International Conference on George Orwell's Nineteen Eighty-Four and its Implication for Life Today, The Ohio State University, Columbus, Ohio, U.S.A., 4-6 May 1984. Appearing under a new title "George Orwell's 1984 and After: A Study in Societal Psychology and the Impact of Technology on Human Values," Ibadan Journal of Humanistic Studies, no. 6 (1986).

_____. "Pascal's Wager and the Atheist's Dilemma," International Journal for Philosophy of Religion, 17, no. 3 (1985):115-129.

_____. "A Philosophical Analysis of the Yoruba Concepts of Ori and Human Destiny," International Studies in Philosophy, XVII, no. 1, (1985):53-69.

_____. "An African Concept of Human Personality: The Yoruba Example," Ultimate Reality and Meaning, 7, no. 3 (1984):189-200.

_____. "God and Social Reality: Reflections on the Pains of Growth," presented at the New Ecumenical Association Conference at Dorado Beach, Puerto Rico, 30 December 1983 to 4 January 1984.

_____. "Immortality of the Soul and the Yoruba Theory of Seven Heavens (Orun Meje)," Journal of Cultures and Ideas, 1, no. 1 (December 1983):31-59.

_____. "Ifa as a Repository of Knowledge." Paper presented at the XVIIth World Congress of Philosophy, Montreal, 21-27 August 21-27 1983. Also in ODU: A Journal of West African Studies, no. 23, (1983):116-121.

_____. "Robin Horton's 'Philosophy': An Outline of Intellectual Error," a paper (105 pp.) presented at the Philosophy Colloquium, Faculty of Arts, University of Ife, 20 June 1978.

_____. "Formal Logic and the Paradox of Excluded Middle." International Logic Review, no. 15 (June 1977):40-52.

_____. "John Stuart Mill's Theory of Logic and Scientific Method as a Rejection of Hume's Scepticism with Regard to the Validity of Inductive Reasoning." Ph.D. diss., University of Toronto, 1974.

Malinowski, Bronislaw. Magic, Science and Religion and Other Essays. New York: Doubleday, 1954.

Markus, R. A. "Augustine, St." in Paul Edwards, ed., The Encyclopedia of Philosophy. 2 vols. New York: Macmillan, 1972.

_____. "St. Augustine" in D. J. O'Conner, ed., A Critical History of Western Philosophy. New York: The Free Press, 1964.

Marx, Karl. The Wisdom of Karl Marx. New York: Philosophical Library, 1967.

Maurier, Henri. "Do We Have an African Philosophy?" in Richard A. Wright, ed., African Philosophy. 3rd ed. New York: University of America Press, 1983.

Mbiti, John S. Concepts of God in Africa. New York: Praeger Publishers, 1970.

_____. African Religions and Philosophy. New York: Doubleday, 1959.

McBriar, A. M. Fabian Socialism and English Politics, 1884-1918. Cambridge: Cambridge University Press, 1962.

Menkiti, Ifeanyi A. "Person and Community in African Traditional Thought," in Richard A. Wright, ed., African Philosophy. 3rd ed. New York: University of America Press, 1983.

Mill, John Stuart. A System of Logic: Ratiocinative and Inductive. 8th ed. London: Longman, 1970.

_____. Utilitarianism. Indianapolis: Bobbs-Merrill, 1957.

Minkus, Helaine K. "Causal Theory in Akwapim Akan Philosophy," in Richard A. Wright, ed., African Philosophy. 3rd ed. New York: University of America Press, 1983.

Moody, J. O. Phylochemical examination of Zanthoxylum (fagara) rubescenes. M. Phil. Thesis, University of Ife, 1980.

Nagel, Ernest. "In Defense of Atheism," in Paul Edwards and Arthur Pap, eds., A Modern Introduction to Philosophy. New York: The Free Press, 1965.

Needham Joseph, and Gwei-Djen, Lu. "Chinese Medicine," in F. N. L. Poynter, ed., Medicine and Culture. London: Wellcome Institute of the History of Medicine, 1969.

Nkrumah, Kwame. Consciencism: Philosophy and Ideology. New York: Monthly Review Press, 1964.

_____. I Speak of Freedom: A Statement of African Ideology. New York: Praeger, 1961.

Nsimbi, M. B. Omweso, A Game People Play in Uganda. Los Angeles: University of California, 1968. (African Studies Center, Occasional Paper.)

Nwankwo, Ezeabazili. African Science; Myth or Reality? New York: Vantage Press, 1978.

Nwanwene, Omoregbe. "Awolowo's Strategy and Tactics of the People's Republic of Nigeria - A Review Article." Quarterly Journal of Administration, V, no. 2 (January 1971).

_____. "Awolowo's Political Philosophy." Quarterly Journal of Administration (Institute of Administration, University of Ife), IV, no. 2 (1970).

Nyerere, Julius. Man and Development. London: Oxford University Press, 1974.

_____. Ujamaa: Essays on Socialism (originally published as a TANU pamphlet in April 1962). London/New York: Oxford University Press, 1971.

_____. Freedom and Socialism: Uhuru na Ujamaa. Dar-e-Salaam: Oxford University Press, 1968.

Ogden C. K., and Richards, I. A. Meaning of Meaning. New York: Harcourt, Brace and World, Inc., 1923.

Ogundimu, B. A. "Personality Variable in Political Leadership and Decision-Making: An Analysis of Obafemi Awolowo's Operational Codes." Quarterly Journal of Administration, XII, no. 3 (April 1978).

Oguah, Benjamin Eruku. "African and Western Philosophy: A Comparative Study," in Richard A. Wright, ed., African Philosophy. 3rd ed. New York: University of America Press, 1983.

Oizerman, Theodor. Problems of the History of Philosophy. Moscow: Progress Publishers, 1973, translated by Robert Daglish.

Olela, Henry. "The African Foundations of Greek Philosophy," in Richard A. Wright, ed., African Philosophy. 3rd ed. New York: University of America Press, 1983.

Omoboriowo, Akin. Awoism: A Select Theme of The Complex Ideology of Chief Obafemi Awolowo. Ibadan: Evans Brothers, 1982.

Onwuanibe, Richard C. "The Human Person and Immortality in IBO (African) Metaphysics," in Richard A. Wright, ed., African Philosophy. 3rd ed. New York: University of America Press, 1983.

Oruka, H. Odera. "Sagacity in African Philosophy," International Philosophical Quarterly (December 1983).

_____. "The Fundamental Principles in the Question of African Philosophy." Second Order, IV, no. 1 (1975):44-45.

_____. "Mythologies as African Philosophy." East African Journal, 9 (October 1972).

Orwell, George. 1984. (Preface by Walter Cronkite). New York: New American Library, 1983.

Parrinder, E. G. African Traditional Religions. London: Hutchinson's University Library, 1962.

Patcher, H. M. Paracelsus: Magic into Science. New York: Collier Books, 1961.

Plato, The Republic, translation by Paul Shorey in Plato's Collected Dialogues, edited by Edith Hamilton and H. Cairns, New York: Bollingen Foundation, 1964.

_____. Apology; The Phaedo in Plato's Collected Dialogues, edited by Edith Hamilton and H. Cairns, New York: Bollingen Foundation, 1964.

Popper, Karl. Objective Knowledge: An Evolutionary Approach. Oxford: Clarendon Press, 1972.

_____. Conjectures and Refutations: The Growth of Scientific Knowledge. New York: Harper & Row, 1963.

_____. The Open Society and Its Enemies (in 2 volumes), London: Routledge and Kegan Paul, 1952.

Poynter, Noel. Medicine and Man. Harmondsworth, Middlesex: Penguin Books, 1971.
Rapoport, Anatol. Operational Philosophy. San Francisco: International Society for General Semantics, 1969.
Robson, J. M. The Improvement of Mankind. Toronto: University of Toronto Press, 1968.
Ruch, E. A. "Is There an African Philosophy?" Second Order, III, no. 2 (1974).
Russell, Bertrand. The Art of Philosophizing. New York: Philosophical Library, 1968.
_____. The Scientific Outlook. New York: W. W. Norton, 1962.
_____. History of Western Philosophy. 18th ed. London: George Allen & Unwin, 1962).
Ryle, Gilbert. Dilemmas. Cambridge: Cambridge University Press, 1966.
Sanda, A. O. "The Scientific or Magical Ways of Knowing: Implications for the Study of African Traditional Healers." Second Order, VII, nos. 1 & 2 (January and July, 1978).
Schell, Jonathan. The Fate of the Earth. New York: Avon Books, 1982.
Senghor, Leopold Sedar. On African Socialism. New York: Praeger, 1964.
_____. Liberte. Paris: Editions de Seuil, 1964.
Smart, J. J. C. Between Science and Philosophy. New York: Random House, 1968.
Snyder, Michael A. "Is a New Dark Age Coming?" The Plain Truth, January 1984.
Sodipo, J. O. "Notes on the Concept of Cause and Chance in Yoruba Traditional Thought," Second Order, II, no. 2 (1973):12-20.
Sofowora, Abayomi. "Man, Plants, and Medicine in Africa." An Inaugural lecture delivered at the University of Ife, Ile-Ife, Nigeria on 7 January 1981.
Somervell, D. C. English Thought in the Nineteenth Century. London: Methuen, 1964.
Stace, W. T. "Metaphysics and Meaning," in Paul Edwards and Arthur Pap, eds., A Modern Introduction to Philosophy. New York: The Free Press, 1965.
Stewart David, and Blocker, H. Gene. Fundamentals of Philosophy. New York: Macmillan, 1982.
Taylor, A. E. Francis Bacon. London: Oxford University Press, 1926.
Tempels, Placide. Bantu Philosophy. Paris: Presence Africaine, 1959.
Thompson, Justice Adewale. African Beliefs: Science or Superstition? Ibadan: Newton House Publications, 1977.
"Traditional Medicine in Zaire: Present and Potential Contribution to the Health Services." Ottawa, Ontario: International Development Research Center, 1980.

Turner, V. W. Ndembu Divination, Its Symbolism and Techniques. Manchester: Manchester University Press, 1961. (Rhodes-Livingston Papers, No. 31.)

Ulam, Adam B. Philosophical Foundation of English Socialism. New York: Octagon Books, 1964.

Vickers, Geoffrey. "Medicine's Contribution to Culture," in F. N. L. Poynter, ed., Medicine and Culture. London: Wellcome Institute of the History of Medicine, 1969.

Von Mises, Richard. Positivism. New York: Dover Publications, 1968.

Waismann, Friedrich. "How I See Philosophy," in A. J. Ayer, ed., Logical Positivism. New York: The Free Press, 1959.

Walker, Benjamin. Encyclopedia of Metaphysical Medicine. London: Routledge & Kegan Paul, 1978.

Wilson, Harold. The Relevance of British Socialism. London: Weindenfield & Nicholson, 1964.

Wiredu, Kwasi. "How Not to Compare African Traditional Thought with Western Thought." Ch'Indaba, no. 2 (July/December 1976). Reprinted in Richard A. Wright, ed., African Philosophy. 3rd ed. New York: University of America Press, 1983.

_____. "Philosophy and African Culture," in Richard A. Wright, ed., African Philosophy. 3rd ed. New York: University of America Press, 1983.

_____. Philosophy and African Cultures. Cambridge: Cambridge University Press, 1980. (Part III).

_____. "On an African Orientation in Philosophy." Second Order, I, no. 1 (1972):3-13.

Whorf, B. L. Language, Thought and Reality. New York, 1956.

Wright, Richard A. "Investigating African Philosophy," in Richard A. Wright, ed., African Philosophy. 3rd ed. New York: University of America Press, 1983.

Zaslavski, C. African Counts: Number and Pattern in African Culture. Boston: Prindle, Weber & Schmidt, 1973.

INDEX

151

Epicurus, 26.
epistemology, 5-9, 21, 23, 27, 33, 34, 39, 41, 48-51, 69-72, 87, 91.
equality, 67, 68, 72, 73, 78.
esusu, 62.
ethnophilosophy, 11-4, 31, 34-9, 44, 54.
euthanasia, 53.
existentialism, 26, 40, 47, 53.

Farrow, S. S., 96, 97.
fatalism, 50.
feminism, 53.
fetish, 6.
Fichte, J. G., 25.
freedom, 6, 41, 47, 50, 66, 67, 75, 82, 85.
Fu Hsi, 104.
Fulbright, xv, xvi, xvii.

gbetugbetu, 95, 97.
gbongbo arun, 89.
Ghana (kingdom), 30.
Gilkey, Langdon, 84.
giri, 101.
God(s), 1, 3, 4, 8, 25, 27, 37, 41, 47, 59, 60, 65-68, 73, 76, 77, 80-3, 87.
Greek philosophy, 1, 11, 51.
griot, 91.

Hallen, Barry, 12.
Hausa, 12.
Hegel, W. F. G., 28, 43, 67.
Hobbes, Thomas, 1, 28, 59, 65.
Horton, Robin, 11, 35, 41.
Hountonji, Paulin, 12, 13.
Hume, David, 12, 28, 51.

Ibo, 12, 14, 49, 50.
idealism, 17, 26.
Identity Theory, 49.
Ifa, xv, 5-10, 41, 50, 51, 87-9, 114-6.
immortality, 1, 4, 5, 24-30, 34, 41, 45, 49, 51, 90.
imperialism, 65, 66, 74.
Indian philosophy, 53, 57, 109.

inductive, 2, 28, 71, 92.
innate ideas, 64.
Islam, 30, 67, 68, 84.
iwosan, 88, 107, 115.

Jahn, Janheinz, 11, 35-7.
James, William, 17, 28, 51.
Jefferson, Thomas, 45.
Jevons, W. S., 9.
Joad, C. M., 24.
Johnson, Samuel, 25.
justice, 59, 67, 78, 84.

Kagame, Alexis, 11, 35, 36.
Kant, Immanuel, 12, 13, 24, 27, 28, 66, 71.
Keita, Lancinay, 30.
Koffi, Niamey, 12.

Lambo, J. A., xv, 105.
Lambo, T. A., xiii, 98, 106.
language, 10-3, 17-20, 25, 41, 42, 47, 53-8, 85, 92, 97.
Laplace, P. S., 8, 9, 41, 51, 81, 88.
Lebeuf, Jean-Paul, 48.
Leibniz, Gottlieb, 28.
Levy-Bruhl, Lucian, 11, 35-7, 41.
Lincoln, Abraham, 82.
Locke, John, 1, 12, 28, 59, 65, 67.
logic, 6, 8, 23, 25-7, 37-48, 53, 71, 85.
Logical Positivism, 26, 40, 47, 49.
love, 76, 77, 80-3.

madarikan, 90, 93.
magic, 6, 9, 10, 88, 91, 92, 96, 106, 115.
magus, 9, 88.
Makinde, D. I., xv.
Makinde, D. K., xiv.
Mali (Kingdom), 30.
mankala, 43.
Maoist, 64.
Marx, Karl, 2, 26, 28, 51, 60-8, 81-6.
materialism, 2-5, 16, 49, 50.

Spinoza, Baruch, 28.
stoic, 74-8.

technology, 3-5, 16, 19, 21,
    53, 65, 83, 84, 98.
Tempels, Placide, 11, 35, 36.
theology, 23.
Thomas, 87.
Thompson, Adewale, 94, 96, 98.
time, 3, 48, 51.
Toure, Sekou, 66.
Towa, M., 12.
transcendental, 26, 49.
tribal, 13, 36, 63.

ujamaa, 60.

violence, 66, 67.
Voltaire, 84.

Waismann, Friedrich, 25-7.
Walker, Benjamin, 106.
wari, 43.
Webb, Beatrice, 2.
will, 69, 79.
Wiredu, Kwasi, 12, 13.
witch, 90-96, 116.
Wittgenstein, Ludwig, 17, 26.
Wood, J. B., 96.
world view, 27, 28.
Wright, Richard, A., 109.

Yoruba, 5-8, 12-4, 26, 34, 42,
    49-51, 59, 65, 86-90, 93,
    94, 98.

Zahan, 87.

ISBN Prefix 0-89680-

25. Kircherr, Eugene C. ABBYSSINIA TO ZIMBABWE: A
Guide to the Political Units of Africa in the
Period 1947-1978. 1979. 3rd ed. 80pp.
100-4 $ 8.00*

27. Fadiman, Jeffrey A. MOUNTAIN WARRIORS: The
Pre-Colonial Meru of Mt. Kenya. 1976. 82pp.
060-1 $ 4.75*

36. Fadiman, Jeffrey A. THE MOMENT OF CONQUEST:
Meru, Kenya, 1907. 1979. 70pp.
081-4 $ 5.50*

37. Wright, Donald R. ORAL TRADITIONS FROM THE
GAMBIA: Volume I, Mandinka Griots. 1979.
176pp.
083-0 $12.00*

38. Wright, Donald R. ORAL TRADITIONS FROM THE
GAMBIA: Volume II, Family Elders. 1980.
200pp.
084-9 $15.00*

39. Reining, Priscilla. CHALLENGING DESERTIFICA-
TION IN WEST AFRICA: Insights from Landsat into
Carrying Capacity, Cultivation and Settlement
Site Identification in Upper Volta and
Niger. 1979. 180pp., illus.
102-0 $12.00*

41. Lindfors, Bernth. MAZUNGUMZO: Interviews with
East African Writers, Publishers, Editors, and
Scholars. 1981. 179pp.
108-X $13.00*

42. Spear, Thomas J. TRADITIONS OF ORIGIN AND
THEIR INTERPRETATION: The Mijikenda of Kenya.
1982. xii, 163pp.
109-8 $13.50*

43. Harik, Elsa M. and Donald G. Schilling. THE
POLITICS OF EDUCATION IN COLONIAL ALGERIA AND
KENYA. 1984. 102pp.
117-9 $11.50*

44. Smith, Daniel R. THE INFLUENCE OF THE FABIAN
COLONIAL BUREAU ON THE INDEPENDENCE MOVEMENT IN
TANGANYIKA. 1985. x, 98pp.
125-X $ 9.00*

45. Keto, C. Tsehloane.   AMERICAN-SOUTH AFRICAN
RELATIONS 1784-1980: Review and Select Biblio-
graphy.   1985.   159pp.
128-4 $11.00*

46. Burness, Don, and Mary-Lou Burness, ed.
WANASEMA: Conversations with African Writers.
1985.   95pp.
129-2 $ 9.00*

47. Switzer, Les.   MEDIA AND DEPENDENCY IN SOUTH
AFRICA: A Case Study of the Press and the
Ciskei "Homeland".   1985.   80pp.
130-6   9.00*

48. Heggoy, Alf Andrew.   THE FRENCH CONQUEST OF
ALGIERS, 1830: An Algerian Oral Tradition.
1986. 101pp.
131-4 $ 9.00*

49. Hart, Ursula Kingsmill.   TWO LADIES OF COLONIAL
ALGERIA: The Lives and Times of Aurelie Picard
and Isabelle Eberhardt.   1987.   156pp.
143-8 $9.00*

50. Voeltz, Richard A.   GERMAN COLONIALISM AND THE
SOUTH WEST AFRICA COMPANY, 1894-1914.   1988.
143pp.
146-2 $10.00*

52. Northrup, David.   BEYOND THE BEND IN THE RIVER:
African Labor in Eastern Zaire, 1865-1940.
1988.   195pp.
151-9 $12.00*

Latin America Series

1. Frei, Eduardo M.   THE MANDATE OF HISTORY AND
CHILE'S FUTURE.   Tr. by Miguel d'Escoto.
Intro. by Thomas Walker.   1977.   79pp.
066-0 $ 8.00*

4. Martz, Mary Jeanne Reid.   THE CENTRAL AMERICAN
SOCCER WAR: Historical Patterns and Internal
Dynamics of OAS Settlement Procedures.   1979.
118pp.
077-6 $ 8.00*

5. Wiarda, Howard J.   CRITICAL ELECTIONS AND
CRITICAL COUPS: State, Society, and the
Military in the Processes of Latin American
Development.   1979.   83pp.
082-2 $ 7.00*

6. Dietz, Henry A., and Richard Moore. POLITICAL PARTICIPATION IN A NON-ELECTORAL SETTING: The Urban Poor in Lima, Peru. 1979. viii, 102pp.
    085-7 $ 9.00*

7. Hopgood, James F. SETTLERS OF BAJAVISTA: Social and Economic Adaptation in a Mexican Squatter Settlement. 1979. xii, 145pp.
    101-2 $11.00*

8. Clayton, Lawrence A. CAULKERS AND CARPENTERS IN A NEW WORLD: The Shipyards of Colonial Guayaquil. 1980. 189pp., illus.
    103-9 $15.00*

9. Tata, Robert J. STRUCTURAL CHANGES IN PUERTO RICO'S ECONOMY: 1947-1976. 1981. xiv, 104pp.
    107-1 $11.75*

10. McCreery, David. DEVELOPMENT AND THE STATE IN REFORMA GUATEMALA, 1871-1885. 1983. viii, 120pp.
    113-6 $ 8.50*

11. O'Shaughnessy, Laura N., and Louis H. Serra. CHURCH AND REVOLUTION IN NICARAGUA. 1986. 118pp.
    126-8 $11.00*

12. Wallace, Brian. OWNERSHIP AND DEVELOPMENT: A Comparison of Domestic and Foreign Investment in Columbian Manufacturing. 1987. 186pp.
    145-4 $12.00*

13. Henderson, James D. CONSERVATIVE THOUGHT IN LATIN AMERICA: The Ideas of Laureano Gomez. 1988. 150pp.
    148-9 $11.00*

14. Summ, G. Harvey, and Tom Kelly. THE GOOD NEIGHBORS: America, Panama, and the 1977 Canal Treaties. 1988. 135pp.
    149-7 $11.00*

Southeast Asia Series

31. Nash, Manning. PEASANT CITIZENS: Politics, Religion, and Modernization in Kelantan, Malaysia. 1974. 181pp.
    018-0 $12.00*

38. Bailey, Conner. BROKER, MEDIATOR, PATRON, AND KINSMAN: An Historical Analysis of Key Leadership Roles in a Rural Malaysian District. 1976. 79pp.
024-5 $7.00*

40. Van der Veur, Paul W. FREEMASONRY IN INDONESIA FROM RADERMACHER TO SOEKANTO, 1762-1961. 1976. 37pp.
026-1 $4.00*

43. Marlay, Ross. POLLUTION AND POLITICS IN THE PHILIPPINES. 1977. 121pp.
029-6 $7.00*

44. Collier, William L., et al. INCOME, EMPLOYMENT AND FOOD SYSTEMS IN JAVANESE COASTAL VILLAGES. 1977. 160pp.
031-8 $10.00*

45. Chew, Sock Foon and MacDougall, John A. FOREVER PLURAL: The Perception and Practice of Inter-Communal Marriage in Singapore. 1977. 61pp.
030-X $6.00*

47. Wessing, Robert. COSMOLOGY AND SOCIAL BEHAVIOR IN A WEST JAVANESE SETTLEMENT. 1978. 200pp.
072-5 $12.00*

48. Willer, Thomas F., ed. SOUTHEAST ASIAN REFERENCES IN THE BRITISH PARLIAMENTARY PAPERS, 1801-1972/73: An Index. 1978. 110pp.
033-4 $ 8.50*

49. Durrenberger, E. Paul. AGRICULTURAL PRODUCTION AND HOUSEHOLD BUDGETS IN A SHAN PEASANT VILLAGE IN NORTHWESTERN THAILAND: A Quantitative Description. 1978. 142pp.
071-7 $9.50*

50. Echauz, Robustiano. SKETCHES OF THE ISLAND OF NEGROS. 1978. 174pp.
070-9 $10.00*

51. Krannich, Ronald L. MAYORS AND MANAGERS IN THAILAND: The Struggle for Political Life in Administrative Settings. 1978. 139pp.
073-3 $ 9.00*

54. Ayal, Eliezar B., ed. THE STUDY OF THAILAND: Analyses of Knowledge, Approaches, and Prospects in Anthropology, Art History, Economics, History and Political Science. 1979. 257pp.
079-2 $13.50*

56. Duiker, William J. VIETNAM SINCE THE FALL OF SAIGON. Second edition, revised and enlarged. 1986. 281pp.
133-0 $12.00*

57. Siregar, Susan Rodgers. ADAT, ISLAM, AND CHRISTIANITY IN A BATAK HOMELAND. 1981. 108pp.
110-1 $10.00*

58. Van Esterik, Penny. COGNITION AND DESIGN PRODUCTION IN BAN CHIANG POTTERY. 1981. 90pp.
078-4 $12.00*

59. Foster, Brian L. COMMERCE AND ETHNIC DIFFERENCES: The Case of the Mons in Thailand. 1982. x, 93pp.
112-8 $10.00*

60. Frederick, William H., and John H. McGlynn. REFLECTIONS ON REBELLION: Stories from the Indonesian Upheavals of 1948 and 1965. 1983. vi, 168pp.
111-X $ 9.00*

61. Cady, John F. CONTACTS WITH BURMA, 1935-1949: A Personal Account. 1983. x, 117pp.
114-4 $ 9.00*

62. Kipp, Rita Smith, and Richard D. Kipp, eds. BEYOND SAMOSIR: Recent Studies of the Batak Peoples of Sumatra. 1983. viii, 155pp.
115-2 $ 9.00*

63. Carstens, Sharon, ed. CULTURAL IDENTITY IN NORTHERN PENINSULAR MALAYSIA. 1986. 91pp.
116-0 $ 9.00*

64. Dardjowidjojo, Soenjono. VOCABULARY BUILDING IN INDONESIAN: An Advanced Reader. 1984. xviii, 256pp.
118-7 $26.00*

65. Errington, J. Joseph. LANGUAGE AND SOCIAL CHANGE IN JAVA: Linguistic Reflexes of Modernization in a Traditional Royal Polity. 1985. xiv, 198pp.
120-9 $12.00*

66. Binh, Tran Tu. THE RED EARTH: A Vietnamese
Memoir of Life on a Colonial Rubber Plantation.
Tr. by John Spragens. Ed. by David Marr.
1985. xii, 98pp.
119-5 $ 9.00*

67. Pane, Armijn. SHACKLES. Tr. by John McGlynn.
Intro. by William H. Frederick. 1985. xvi,
108pp.
122-5 $ 9.00*

68. Syukri, Ibrahim. HISTORY OF THE MALAY KINGDOM
OF PATANI. Tr. by Conner Bailey and John N.
Miksic. 1985. xx, 98pp.
123-3 $10.50*

69. Keeler, Ward. JAVANESE: A Cultural Approach.
1984. xxxvi, 523pp.
121-7 $18.00*

70. Wilson, Constance M., and Lucien M. Hanks.
BURMA-THAILAND FRONTIER OVERSIXTEEN DECADES:
Three Descriptive Documents. 1985. x, 128pp.
124-1 $10.50*

71. Thomas, Lynn L., and Franz von Benda-Beckmann,
eds. CHANGE AND CONTINUITY IN MINANGKABAU:
Local, Regional, and Historical Perspectives on
West Sumatra. 1986. 363pp.
127-6 $14.00*

72. Reid, Anthony, and Oki Akira, eds. THE
JAPANESE EXPERIENCE IN INDONESIA: Selected
Memoirs of 1942-1945. 1986. 411pp., 20 illus.
132-2 $18.00*

73. Smirenskaia, Ahanna D. PEASANTS IN ASIA:
Social Consciousness and Social Struggle. Tr.
by Michael J. Buckley. 1987. 248pp.
134-9 $12.50

74. McArthur, M.S.H. REPORT ON BRUNEI IN 1904. Ed.
by A.V.M. Horton. 1987. 304pp.
135-7 $13.50

75. Lockard, Craig Alan. FROM KAMPUNG TO CITY. A
Social History of Kuching Malaysia 1820-1970.
1987. 311pp.
136-5 $14.00*

76. McGinn, Richard. STUDIES IN AUSTRONESIAN
LINGUISTICS. 1988. 492pp.
137-3 $18.50*

78. Chew, Sock Foon. ETHNICITY AND NATIONALITY IN
    SINGAPORE. 1987. 229pp.
    139-X $12.50*

79. Walton, Susan Pratt. MODE IN JAVANESE MUSIC.
    1987. 279pp.
    144-6 $12.00*

80. Nguyen Anh Tuan. SOUTH VIETNAM TRIAL AND
    EXPERIENCE: A Challenge for Development. 1987.
    482pp.
    141-1 $15.00*

81. Van der Veur, Paul W., ed. TOWARD A GLORIOUS
    INDONESIA: Reminiscences and Observations of
    Dr. Soetomo. 1987. 367pp.
    142-X $13.50*

82. Spores, John C. RUNNING AMOK: An Historical
    Inquiry. 1988. 190pp.
    140-3 $13.00*

ORDERING INFORMATION

Orders for titles in the Monographs in Inter-
national Studies series should be placed through the
Ohio University Press/Scott Quadrangle/Athens, Ohio
45701-2979. Individuals must remit pre-payment via
check, VISA, MasterCard, CHOICE, or American
Express. Individuals ordering from the United
Kingdom, Continental Europe, Middle East, and Africa
should order through Academic and University
Publishers Group, 1 Gower Street, London WC1E 6HA,
England. Other individuals ordering from outside of
the U.S., please remit in U.S. funds by either
International Money Order or check drawn on a U.S.
bank. Postage and handling is $2.00 for the first
book and $.50 for each additional book. Prices and
availability are subject to change without notice.